FREEDOM FROM ENVIRONMENTAL SENSITIVITIES

By

D. S. NAMBUDRIPAD, M.D., D.C., L.Ac., Ph.D.

Author of

SAY GOOD-BYE TO ILLNESS AND

SAY GOODBYE TO ... SERIES

This book has Revolutionized
The Practice of Environmental Medicine

The doctor of the future will give no medicine,
But will interest his patients
In the care of the human frame, in diet,
And in the cause and prevention of disease.
                                        Thomas A. Edison

Published by
DELTA PUBLISHING COMPANY
6714 Beach Blvd.
Buena Park, CA 90621
(888) 890-0670, (714) 523-8900 Fax: (714) 523-3068
Web site: www.naet.com

DEDICATION

This book is dedicated to
environmentally ill patients, and
anyone wishing to get well.

First Edition, 2005

Copyright © 2005 by D. S. Nambudripad
M.D., D.C., L.Ac., Ph.D.
Buena Park, California
All rights reserved.

Library of Congress Control No: 2004108267

ISBN: 0-9743915-8-1

Printed in U.S.A.

The medical information and procedures contained in this book are not intended as a substitute for consulting your physician. Any attempt to diagnose and treat an illness using the information in this book should come under the direction of an NAET physician who is familiar with this technique. Because there is always some risk involved in any medical treatment or procedure, the publisher and author are not responsible for any adverse effects or consequences resulting from the use of any of the suggestions or procedures in this book. Please do not use this book if you are unwilling to assume the risks. All matters regarding your health should be supervised by a qualified medical professional.

CONTENTS

Freedom From Environmental Sensitivities

Other Books by Dr. Nambudripad include:

Say Good-bye to Illness, 3rd Edition in English

Say Good-bye to Illness, 1st Edition in French

Say Good-bye to Illness, 1st Edition in Spanish

Say Good-bye to Illness, 1st Edition in German

Say Good-bye to Illness, 1st Edition in Japanese

Say Good-bye to Allergy-related Autism, 2nd Edition

Say Good-bye to ADD and ADHD

Say Good-bye to Children's Allergies

Say Good-bye to Your Allergies

Say Good-bye to Asthma

Living Pain Free

The NAET Guide Book, 6th Edition

For details, please visit our web site: www.naet.com

ACKNOWLEDGMENTS

I am deeply grateful to my dear husband, Kris K. Nambudripad, for his inspiration, encouragement and assistance in my schooling and later, in the formulation of this project. Without his cooperation in researching reference work, revision of manuscripts, word processing and proofreading, it is doubtful whether this book would ever have been completed. My sincere thanks also go to the many clients who have entrusted their care to me, for without them I would have had no case studies, no technique and certainly no extensive source of personal research upon which to base this book.

I am also deeply grateful to my patients Meg B., Shirley R., Michael, Toby W. , Joyce B., Karen W., Lonnie Y., Dottie J., Vickie V., Martha E., Jean E., Weber G., Cathy C., Kathy K., Lavern B., Garcia I., Steven M., Robert B., Helen T., George L., Jamie B., Janet J., Bob L., Gary M., John M., James L., Chuck B., Sharon B., Ann B., Lettie V., Amy, and many other hundreds of my beloved environmental sensitivity patients who worked with me and are finally living normally now.

I also would like to express my heartfelt thanks to my friends who are excellent NAET practitioners and supported me by providing case studies, patients' testimonials, and constant encouragements to bring this book out. Without their ardent help, the writing of this book would have been only a dream.

Additionally, I wish to thank Mala and Mohan Moosad for allowing me to work on the book by relieving me from the clinic duties, and helping me in the formulation of this book. I do not have enough words to express my thanks to Dr. Prince, Dr. Rish Namboodiri, Doretta Zemp, M.A., M.S., M.F.T., and many of my friends who wish to remain anonymous for proofreading the work, and Sridharan at Delta Publishing for his printing expertise. I am deeply grateful for my professional training and the knowledge and skills acquired in classes and seminars on chiropractic and applied kinesiology at the Los Angeles College of Chiropractic in Whittier, California; the California Acupuncture College in Los Angeles; SAMRA University of Oriental Medicine, Los Angeles; University of Health Sciences, Antigua, and the clinical experience obtained at the clinics attached to these professional schools.

I extend my sincere thanks to my great teachers. They have helped me to grow immensely at all levels. My professional mentors are also indirectly responsible for the improvement of my personal health as well as that of my family, patients and other NAET practitioners among whom also include are countless doctors of Western and Oriental medicine, chiropractic, osteopathy, allopathy, as well as their patients.

Many of the nutritionists instrumental in this process (the developing NAET) were professors at the institutions that I have mentioned earlier. Their willingness to completely dedicate to teaching, as well as their commitment of personal time to give the interviews necessary to complete this work, places them beyond my mere expressions of gratitude. They are living testaments to the greatest ideals of the medical profession.

Dr. Devi S. Nambudripad
Los Angeles, CA

FOREWORD

By

Sandra C. Denton, M.D.

I first became aware of the scope of environmental aller
gens through membership and training in the American
Academy of Environmental Medicine. Before that, like
many of you, I only considered common inhalants such as pollen,
grass, trees, animals, etc to be the cause of allergy symptoms.
Even my concept of what comprised allergy symptoms, such as
watery eyes, runny nose, sneezing, and asthma was far too limited.
I had no idea of the vastness or the extent that our environment
(work, hobby, clothes, temperature, pesticides, even the food container) might have on our physical and mental well-being. Through
experience with AAEM I began to look at things differently. Then
I learned NST and NAET and my detective skills at identifying
hidden environmental allergens greatly increased along with my
treatment success rate.

One patient came to me suffering from severe suicidal depression since her divorce a few years earlier. She had been to
many counselors to no avail. It was determined that after her divorce she had moved into a brand new apartment complex and a
new office with new carpet and paint. Every 6 months her office
was remodeled with additional exposure. Her athletic club had
installed new carpet and she was always worse after her work-
outs, especially when doing floor exercises (sit-ups, push-ups, etc.).
It was finally determined that the cause of the depression was

formaldehyde and following treatment, she was totally free of depression and has remained healthy for several years now.

Another interesting case involves a five-year-old girl who a year earlier had hit her head after a fall on the playground. She was unconscious for about five minutes and remained very sluggish. An MRI was done at the hospital which was negative. From that time on the little girl became extremely allergic to anything electrical, even underground electrical fields, with grand mal seizures. The parents had been to numerous doctors, including Dr. Bill Rea's environmental clinic in Texas. No one had been able to help. The parents even had the electricity to their house disconnected and used candles and propane lamps and heater. The child did better, but if she ran in the backyard and got too close to the underground wires she would experience a seizure. The symptoms were so severe that she would experience a seizure if the government was doing electrical experiments at Elmendorf Air Force Base which was across town from their home. The family had been advised to sell their home and move from Alaska to the desert of New Mexico.

A friend of the family implored me to intervene on their behalf. The child could not even come into the office because of the fluorescent lights and electrical energy of the building. I went to the house and through NST (Neuromuscular sensitivity Testing) determined the child to be allergic to vitamin C, magnesium (which she was being given to control the seizures), radiation and iron. I remembered as a child having a toy that contained iron filings and a magnetic pencil to draw pictures. When finished, you shake the toy and get a blank screen and draw a new picture. It occurred to me that after the accident, she might have had a bruise in the brain (hematoma-a collection of red blood cells which contain iron). The effect of the MRI (Magnetic Resonance Imaging) in the presence of iron may have damaged her brain making her "super sensitive" and translating into seizures around electricity. After treating the four allergic items with NAET, the child is totally well and can go anywhere she likes seizure free. Their life has been restored to normal through the power of NAET.

Convinced that their son was just lazy, Johnny's parents didn't mention that he was not doing well with his home schooling. They brought him to me because he was constantly getting sick with swollen lymph glands and sinus problems. Then one day they told me that 15-30 minutes after beginning his lessons, he couldn't concentrate. Then he would go play and he seemed fine. When I heard this, I immediately checked him on his textbook, which he had with him. He was so allergic to the color pictures and ink on the book that he would almost pass out. Two days after NAET treatment he sat down for three straight hours without a break and finished his lesson and has done well ever since. If the allergy to ink had not been identified and treated, he might never have finished school. His immune system was healed by treating several other environmental allergens and he rarely gets an infection anymore. All of this was accomplished without needles for testing or shots for treatment, much to the relief of this patient.

No longer is it necessary to live in a "bubble," or be paralyzed by multiple chemical sensitivities and a prisoner in your own home, or a social outcast because no one can come visit you. NAET has revolutionized my practice. Both my patients and I are grateful for both the simplicity and the RESULTS of this remarkable treatment.

It is sometimes necessary to collect an "air sample" of the patient's home or work/school environment and treat with NAET to successfully treat the combination of multiple things in the air (mold, dust, cleaning chemicals, insecticides, animal dander, etc.) that may be causing a problem. This has been particularly effective in cases where symptoms remained even after clearing the individual items.

Freedom From Environmental Sensitivities

Thank you again Dr. Devi! What a blessing not to have to give up your favorite pet, hobby or occupation or spend half of your earnings on medication with all of its side-effects.

Sandra C. Denton, M.D.
Alaska Alternative Medicine Clinic, LLC
3333 Denali St
Anchorage, AK 99503
907-563-6200

Preface

Since childhood I suffered from a multitude of health problems. Because of this prolonged and firsthand experience with ill health, I became focused on health-related problems, particularly those related to allergies; this, in turn, resulted in my natural inclination to pursue medicine as a profession. Consequently, I became a registered nurse, chiropractor, kinesiologist, acupuncturist, earned a doctorate in philosophy in Oriental medicine and, now an M.D. I have been in school all my life, searching for answers to help myself and others with health-related disorders. I began specializing in the treatment of the allergic patient, using methods I learned through an intensive study of Oriental medicine combined with the more traditional Western methods learned in various schools of Western medicine.

During my studies and early practice as an allergist, while using eclectic methods of treatment procedures, I discovered a technique that eliminated most of my health problems.

Integrating the relevant techniques from the various fields I studied, combined with my own discoveries, this new treatment has become the focus of my practice. There is no known successful method of treatment for environmental sensitivities using Western medicine except avoidance, which means deprivation and frustration. Each of the disciplines I studied provided bits of knowledge that I used in developing this new treatment to permanently eliminate allergies and sensitivities to the treated allergen, which is now known throughout the world as Nambudripad's Allergy Elimination Techniques, or NAET for short.

I was an example of environmental sensitivities, chemical sensitivities, I was a universal reactor before treated and eliminated my allergies and sensitivities. I reacted to everything under the sun, even sun by radiation. As an infant living in India, my birthplace, I had severe infantile eczema, which lasted until I was seven or eight years old. My eczema started at 11 months old after eating a whole tomato one night instead of dinner, as my mother recalled. After that, I was administered Western medicine, Ayurvedic herbal medicine, various cleansing diets without a break until I was eight years old. I was born and brought up as a vegetarian. From childhood, my major diet was organically grown vegetables–fruits and grains grown in our fields (without using any artificial or chemical fertilizers). My family was into farming. We grew everything in our farm including coffee and tea. We had cows, goats and buffulos and we made our own cheese, butter and yogurt without any additive or colorings. We made coconut oils and sesame oils at home. We had to buy salt and kerosine from outside (In my village there was no electricity then. We used kerosine lamps for lighting). Rest everything we had made at home.

When I turned one year old, after the 'Tomato' dinner my health problems continued one after another: It started as generalized skin rashes, hives, eczema, dermatitis, sinusitis, sinus troubles, angioneurotic edema (swelling in different body parts– frequently eyelids, face, sometimes even throat without any warning or known cause), and severe arthritis. My saga continued... frequent colds and fevers, flu-like symptoms, constant post nasal drips, thick mucus in the throat, pain and swelling in the joints, severe fatigue, and general body ache. By the time I was ten, I began having severe migraines, then later, severe PMS. I never had anything mild. My symptoms were always in extremes making living difficult. I spend my childhood visiting doctors and taking medicines, yet I still suffered constantly.

After discovering NAET, treating myself and eliminating known allergens, years later, pesticides still remained a big problem for me. I continued to treat each pesticide as it crossed my path. After treatment, the particular type did not bother me anymore. But I often wondered why I was so allergic to pesticides.

The mystery was solved when as an adult I visited my birthplace in Kerala, South India, in 1995. I decided to take a walk along the valley along with my cousins where we played as children, revisiting and refreshing the nostalgic memories of our old times. After we walked for about five minutes, one of my cousins looked at my face and let out a scream. My face had turned red, with huge hives all over my face and neck. My body was hot. I felt my legs go very weak, as if they would give up any minute. My lungs were working hard to get air. My cousins were scared, not knowing what was happening to me. I, too was surprised at this sudden, unexpected reaction when I thought I had conquered all my allergies. We turned back. I leaned onto my cousin's shoulder to prevent falling.

Suddenly, I noticed some white patches on all the leaves and grasses on the ground, as if someone had sprayed diluted white paint all over.

"What is this?" I asked, pointing toward the white patches.

One of my cousins answered, "It is DDT. We have had some problems with mosquitoes lately, due to the new rubber plantations. The health department has been spraying DDT for months now every week. They just sprayed this morning."

I sighed in relief. So again, pesticide-DDT was a culprit. I asked my cousin to collect some of the leaves. Once I reached the house, I quickly treated myself through NAET for DDT along with the leaves. In less than 10 minutes, my breathing became normal. In 40 minutes, all the hives disappeared. My body strength and leg strength returned. I became normal again. Thank you God, for granting me NAET.

I then remembered my young days in the valley. The cause of my illness could have been the DDT spray. I always felt fatigued and listless while I was there. It was a chore for me to get through the days, a chore to breathe and a chore to live. I felt much better healthwise when I moved away to another city to continue with my school.

After years of NAET, I do not suffer shortness of breath anymore, but I still suffer coughing spells with any irritation to my lungs. One of my recent lung scans revealed that I have congenital-emphysematous lungs. My pulmonologist couldn't believe that in spite of having such lungs, I still do very well; my respiratory tract works normally. After enjoying complete cure from my respiratory problems for 20 years, I was very surprised to see the scar of my childhood illness is still seen in the lung scan. Usually minor scars diminish over time and in a few years there won't be any trace left. My lung problem could have been real big, otherwise after 20 years, the scar will not be visible now.

NAET is a simple but powerful technique that involves art and science from various medical disciplines. It can give relief from multitudes of commonly seen health disorders, those originating from some allergic reactions. It requires patience on the part of the practitioner as well as the patient because it sometimes takes numerous office visits for optimum recovery, especially recovery from long-term or serious health disorders.

NAET is a very effective treatment, that some people can experience immediate relief from their symptoms. Everyone should be treated for the "Basic-15 allergen groups" first. During my 21 years of practice, I have seen all variations. Some patients with good immune system have taken three or four NAET treatments and lived happily ever after, whereas some others who have poor immune systems, a long list of sicknesses in the family tree, and poor general health habits, can take hundreds of NAET treatments to straighten out their lives. Chronic suffering, physical, physiological or mental abuse of any kind can cause tissue damage within the body and body organs. People falling in this group take a longer time to recover.

We continue to expose ourselves to harmful allergens many times every day. Continuous exposures and re-exposures to various allergens throughout one's daily living can aggravate any health condition. In ailing people, presence of any allergen can trigger immediate allergic reactions, usually attacking the weakest organ first. Then eventually, the other organs will be affected. For example: a person with chronic bronchitis or asthma usually has weak, less efficient, hypersensitive lungs. In this person, lungs are the weak area in the body. Whenever this person comes in contact with an allergen his/her lungs will begin to react, producing symptoms like postnasal drip, mucus production in the throat, cough, sinusitis or asthma (the common pathological symptoms of the lungs). To get the asthma or chronic bronchitis or sinusitis under control, one may have to eliminate allergies to many allergens from his/her food and environment. But the good news is this: They too can benefit from NAET even though they may have to take a series of treatments.

Some may wonder why people get more allergies in this century than did our ancestors. Evolution is the probable cause for this. Our ancestors had different health problems in their days, like bacterial infections, parasitic infestations, etc. There weren't as many chemicals or environmental hazards in those days. Their genes were never exposed to the thousands of man-made chemicals that we encounter today which trigger unlimited adverse reactions.

Genes have the innate ability to adapt to new environment, substances, or life-style, if nurtured in healthy environments. People who were born to healthy parents, and lived in clean and healthy environments have less sickness. When people were born to unhealthy, or abusive parents, exposed to toxic chemicals, pesticides, fumes and other toxins in their early days, especially if they continue to live in unhealthy conditions, continue to become victims of physical, physiological or mental abuse they might suffer untold health problems.

That is where NAET comes in. NAET is capable of reprogramming the confused brain by erasing wrongly imprinted messages or by waking up the part of the genes that was dormant and re-imprint correct information about the harmlessness of an environmental toxin, a new chemical, product or life-style. Most physical, physiological and psychological abuses can be easily addressed by NAET and the body can be restored to enjoy the world and worldly things once again.

Chemically and environmentally sensitive people need to be handled gently. Their NAET treatments should be given in the correct order so as to prevent excess stress, so that their already stressed-out immune system will be calmer, confident to heal back to normalcy.

Environmentally or chemically sensitive people have very fragile emotional centers in the brain. Any little incident can cause an emotional allergic response. Their emotions and sensitivities need constant balancing until they heal completely. NAET treatments are designed to balance their physical, physiological and psychological health. If their emotional health is severely affected, it is even beneficial if they can use the help of some psychiatric drugs to keep their symptoms under control so that NAET treatments can be completed without much interruption.

Many people are becoming chemically and environmentally sensitive lately. Doctors and patients are equally frustrated. Some patients live in a bubble-like environment. When one is chemically and environmentally allergic to almost everything, that person's brain-software is completely disrupted. This type of person has completely confused genes. It might take numerous, continuous treatments with NAET to reintroduce every item around that person or replace all incorrect informations in the person's brain with appropriate messages about his/her living environment. Again the good news is this: When they complete the needed NAET treatments, they too can live a normal life among normal people, among the assorted pollutants, enjoying the benefits of the scientific advancements of the 21st Century. NAET is the treatment of 21st Century. I have searched through all available literature in order to see if there is any other type of treatment that can eliminate allergies permanently to the treated allergens like NAET does. So far I have not found any treatment that even comes close.

Environmentally sensitive patients can live like other normal persons, provided they get treated for their allergies by doctors well trained in NAET.

I would feel gratified, indeed, if the up-to-date material compiled herein were to contribute to my readers achieving, maintaining and enjoying good health. I would feel equally gratified if all other healing professions would learn NAET methods and incorporate into their existing treatment procedures in order to help their patients maintain and enjoy better health.

I have tried to focus on the issues of environmentally sick people in this book to let them know that they are not alone in this life saga. I have also tried to inform them of certain means to look into their problems and detect the roots. Once detected, it is easier to remove through NAET in order to lead a normal life. I have published many books on allergies and NAET. Most of my books carry descriptions of the specifics related to a particular health problem. My book *Say Good-bye to Illness* is a concise melting pot for all health-related information. In order to make this book– *Freedom From Environmental Sensitivities* reader-friendly and not repeat materials already found in my previous books, you will not find everything related to NAET herein. For further readings on NAET, please refer to the web site:WWW.naet.com.

If you wish to learn more about acupuncture meridians and possible pathological symptoms, or to find out detailed information on diagnostic testing and evaluations to detect allergies, you might want to read *Say Good-bye to Illness*. Or to further enhance your understanding, you might pursue some of the other relevant books and articles listed in the bibliography at the conclusion of this book.

Stay Allergy-Free and Enjoy Better HEALTH!

Devi S. Nambudripad,
M.D., D.C., L.Ac., Ph.D.
Buena Park, California
July, 2005

INTRODUCTION

WHAT IS NAET?

NAET is an acronym for Nambudripad's Allergy Elimina tion Techniques. Various effective parts of healing techniques from different disciplines of medicine (allopathy, acupuncture, chiropractic, kinesiology and nutrition) have been compiled together to create NAET, to permanently eliminate allergies of all kinds (food and environmental allergies and reactions in varying degrees from mild to severe to anaphylaxis) from the body. NAET is a completely natural, non-invasive and painless holistic treatment. Over 8,000 licensed medical practitioners have been trained in NAET procedures worldwide. Please look up the NAET website, **"www.naet.com"** for more information on NAET and a NAET practitioner near you.

WHAT IS AN ALLERGY?

A condition of unusual sensitivity of one individual to one or more substances (may be inhaled, swallowed or contacted by the skin), which may be harmless or even beneficial to the majority of other individuals. In sensitive individuals, contact with these substances (allergens) can produce a variety of symptoms in varying degrees ranging from slight ADHD to Autism, mild itching to swelling of the tissues and organs, mild runny nose to severe asthmatic attacks, general tiredness or fatigue to severe anaphylaxis. The ingested, inhaled, injected or contacted allergen is capable of alerting the immune system of the body. The frightened and confused immune system then commands the white blood cells to produce immunoglobulins to stimulate the release of neuro-chemical defense forces like histamines from the mast cells. These chemical mediators are released as part of the body's immune response.

WHAT CAUSES ALLERGIES?

- Heredity - inherited from parents, grandparents, ancestors, etc.

- Toxins - produced in the body from: food interactions, unsuitable proteins, bacterial or viral infections; molds, yeast, fungus or parasitic infestation; vaccinations and immunizations; drug reactions; constant contacts with certain irritants like mercury, lead, copper, chemicals, etc.

- Low immune system function - due to surgeries, chronic illnesses, injury, long term starvation, etc.

- Radiation - excessive exposure to television, computers, sun, radioactive materials, etc.

- Emotional factors.

WHAT ARE SOME COMMON ALLERGENS?

- Inhalants: grasses, pollens, flowers, perfume, dust, paint, formaldehyde, smoke, fumes, pollution, etc.

- Ingestants: food, drinks, vitamins, drugs, food additives, etc.

- Contactants: fabrics, chemicals, cosmetics, furniture, utensils, etc.

- Injectants: insect bites, stings, injectable drugs, vaccines, immunization, etc.

- Infectants: viruses, bacteria, contact with infected persons, etc.

- Physical Agents: heat, cold, humidity, dampness, fog, wind, dryness, sunlight, sound, etc.

- Genetic Factors: inherited illnesses or tendency from parents, grandparents, aunts, uncles, great grandparents, etc.

- Molds and Fungi: molds, yeast, candida, parasites, etc.

- Emotional Factors: painful memories of various incidents from past and present.

HOW DO I KNOW IF I HAVE ALLERGIES?

If you experiences any allergic symptoms or unusual physical, physiological or emotional symptoms in the presence of any of the above listed allergens, you can suspect an allergy contributing towards such changes.

WHO SHOULD USE THIS BOOK?

Anyone with an allergy, or anyone with an unsolved health problem should read this book. Your problem may be due to undiagnosed allergies.

HOW IS THIS BOOK ORGANIZED?

Chapter 1 - begins with an unusual case study of a common health problem that seen among people today. This was the first time, the author made the discovery that even a stoke can happen from an allergy.

Chapter 2 - categories of Allergens - describes the various categories of allergens and how they can mimic different symptoms.

Chapter 3 - detecting allergies - explains some commonly used allergy screening measures.

Chapter 4 - order of NAET treatments

Chapter 5 - symptoms of meridians - describes the normal functions of the meridians and how energy interferences create abnormal responses in the meridians leading to allergies and allergy-related health disorders.

Chapter 6 - NAET Testing Procedures - explains neuromuscular sensitivity testing (NST), the effective, noninvasive, easy-to-perform NAET screening procedure to detect allergies.

Chapter 7 - NAET Home-help - describes and illustrates self-balancing techniques and the use of acupressure techniques to give the reader more control over his/her emergency reactions to environmental sensitivities.

The glossary will help you to understand the appropriate meanings of the medical terminology used in certain parts of the book.

The resource guide is provided to assist you in finding products and consultants to support you in dealing with allergies.

The Bibliography covers numerous sources of information on allergies.

A detailed index is included to help you locate your area of interest quickly and easily.

NAET Testimonials

Everyone is exposed to environmental chemicals over which they have no control. Buildings are cleaned with many different chemicals and sprayed with pesticides that are toxic to most people. Scented personal care products, laundry products, and room deodorizers further contribute to the problem. The outgassing of building materials, and newly manufactured products emit many chemicals, particularly formaldehyde. Factory emissions, oil refineries, and automobile exhaust pollute the air. Lawn care products and chemicals used in farming find their way into our food supply. Even with careful planning and lifestyle, people cannot avoid the many exposures to environmental chemicals.

Many people are sensitive to these environmental chemicals and experience myriad symptoms, varying in severity. They cannot live a life even approaching normal without some type of treatment. NAET treatments enable these people to be desensitized to the chemicals so that exposure does not trigger symptoms. While everyone should endeavor to limit chemical exposures in order not to continue to bioaccumulate chemicals in their tissues, NAET allows people to carry out daily and work activities. Gone is the necessity of a remote pristine existence in order to survive.

Frances Taylor, M.A & Jaqueline Krohn, M.D.
Authors of *Allergy Relief and Prevention*
The whole way to natural detoxification
Albuquerque, NM

In our office it is routine for us to see environmental sensitivities go away and stay gone with NAET. The technique is gentle but effective and the results often amaze me and our clients. I am very grateful to Dr. Devi for her work.

Laurie Teitelbaum- Nutritionist, NP
Annapolis, MD. www.EndFatigue.com

Environmental Illness is a complex disorder that can be all consuming and even life shattering for the person suffering with it. Seven years ago after healing my own extreme case of environmental illness, I shifted the focus of my clinical practice to assisting others with this condition. NAET has been a most valuable aid in this regard. For some, NAET alone has led to dramatic improvement. For others, NAET has provided relief and a stability that has enabled the individual to pursue other processes essential for their full recovery.

Thank you Dr. Nambudripad for your wonderful discovery!

Robert Sampson, MD
(Co-author *Breaking Out of Environmental Illness*)
Billerica, MA

After practicing NAET for several years I can truly say that Dr. Nambudripad has formulated a method for handling the majority of illnesses plaguing mankind. We have seen miracles occur on a daily basis in our office and continue to be amazed by this powerful technique. We have had great results taking care of both children and adults suffering from conditions ranging from food allergies to environmental allergies and everything in between. Kids suffering from eczema from food sensitivities now have clear skin. People not able to enjoy their life due to their health conditions could now engage in all the activities they want to without the fear that they will have acute reactions. NAET is truly revolutionary and its success is far reaching. For those of you reading this book praying for relief for your ailments, you may have just found your answer...for the only thing you have to lose is your illness! Thank you Dr. Devi!

Dr. Steve Popkin, D.C.
Ft. Lauderdale, Florida

NAET treatments in my practice have been a blessing to many of my patients and have been rewarding to me. NAET is used successfully in removing environmental allergies causing severe infantile eczema which otherwise would have to be treated with steroids.

James Christianson, D.O.
Coffeyville, KS

1

Discovery of an Allergen that Caused Stroke in a Patient

Living is difficult when one cannot live side by side with environmental factors in this world. People every where are exposed to various environmental factors over which they have no control. The ground one walks about, ground coverings, trees, woods, rocks, sand, stones, plants, flowers, perfume, building materials, furniture, carpet, ceramic tile, different types of metals, utensils, clothes, plastic wares, animals, animal droppings, bees, spiders, other insects, various microorganisms, wind, cold, heat, sun radiation, other radiations, electric and electronic equipments, pesticides, fertilizers, the toxins emitting from any of the above are all naturally seen environmental factors and we encounter most of them in the course of daily living. Many people have been found intolerant to many of these factors. They get varied health discomforts and disorders when they come near these agents. To make matters complicated, our scientists have generated thousands of different chemicals to give color and attraction to the above natural environmental factors. Buildings are treated with formaldehyde then painted with different types of paints, wood is treated with pesticides and other poisons to keep the pests away, grasses and trees are sprayed with strong pesticides, clothes are treated with chemicals, personal care products are not only treated with chemicals but scented with heavy perfume to mask the chemical smells, various laundry products, cleaning agents, room deodorizers all are toxic materials to sensitive people. Living in the crowded cities, near factories, near oil refineries, near parks, near farms (more pesticide sprays), is made difficult to environmentally sensitive people. The outgassing of building materials, especially formaldehyde in the new buildings and radon in the old buildings make life impossible even inside the closed doors of the house. However careful one is to avoid the above listed environmental toxins, life is difficult for the environmentally sensitive individual.

The number of environmentally sensitive people are increasing daily (Rapp, Our toxic world, 2003). The actual fact is that there is hardly any human who is not sensitive to some of the environmental factors. The sensitivity towards these factors are causing different types of reactions and health problems in people too, whether they realize or not. Potentially, one can be allergic to anything one comes in contact with and it can produce any type of disorder or illness in anyone.

Working for the past two decades with various types of allergies and allergic reactions and seeing the full recovery from their "incurable" long term health problems when the allergies are eliminated, I am convinced that just about any illness (other than injuries) can have an involvement of an allergic nature. I keep my mind completely open towards this concept that any health disorder can begin with an allergy. But even with this concept, I was taken by surprise when I discovered the cause of "Sara's STROKE" was due to an allergy–an allergy to the wood cabinet in her room. When the allergy was desensitized with NAET, she completely recovered from her stroke.

The following is the story of a real person who was very fortunate in overcoming a major incident of *Cerebrovascular disorder or stroke* with the help of NAET that usually leaves someone disabled for life. Medical professionals all over the world see many such cases daily; many health disorders appear suddenly without any obvious reason; at least, medical professionals have little clue why certain health problems appear. Most of the commonly performed standard medical diagnostic procedures are not sensitive enough to pinpoint the unusual etiology behind certain health problems. The physician who is disciplined in the systematic methods of diagnostic evaluation taught in medical schools is not usually able to change directions and explore alternate causes other than what they've learned.

The facts of the following case history are quite common. Sara was lucky to have found someone to veer from traditional diagnostic means and explore different avenues for such a common health problem as "stroke." Not too many patients will get a chance to recover 95 percent of body functions from a stroke in 14 days, let alone discover the cause for the stroke. In Sara's case, the major underlying cause for her stroke was environmental and the allergen was discovered quite unexpectedly. She wrote her own story on how she experienced it.

My name is Sara and I have a success story about my health to share with you. A-72year-old woman, I have been fairly healthy all my life until I began getting an early morning headache almost daily and frequent flu's beginning November of 2000. Every couple of weeks I had an attack of 'flu' which usually lasted for a week. My internist had to prescribe antibiotics a couple of times. I also noticed that my head felt heavy and slightly achy if I took a nap in the afternoon in my bedroom. So I began taking the naps in the family room. I had no ill effects from the afternoon naps. The following January, I saw my internist and told him about the headaches and other ailments. He examined me thoroughly and found every test normal except for my diabetes of six-year duration which had been under control until three months ago when my fasting sugar jumped up and remained over 250 mgs/dl. My internist was puzzled and increased my medications and put me on a very strict diet consisting of no salt, no sugar, no fat, but a high protein. He also prescribed antibiotics since my latest 'flu' had turned into a kind of bronchitis. I took antibiotics twice, each session lasting for ten days, with no relief. The doctor was baffled.

On Tuesday, February 13th, 2001, I woke up at about 12:45 a.m. with a severe headache on the right side and back of the head radiating to the top of the head, abdominal bloating, nausea, and vomiting. I became extremely weak from continuous vomiting. Tired and weary I slept for a couple of hours. Waking up at about 10:00 a.m., I noticed that I couldn't get up from the bed. I felt weakness on my whole right side. My daughter helped me sit up. I suddenly felt a cold chill pass through my right side and following that my right hand and right leg, right side of the face, neck and head became numb. My speech became slurred. My daughter thought I had a mild stroke while I was still sitting in the bed talking to her.

I was taken to the hospital immediately and admitted to the special care unit.

As the hours passed by, in the hospital my symptoms got worse. Three hours after reaching the hospital, I could no longer move the right side of my body on my own. My speech was not understandable and I felt my words trapped inside my throat. I was also told that I had redness and mild rashes over my right arm, face and neck. I struggled to get air into my lungs.

I felt a ball sitting in my throat. I couldn't swallow. I began drooling. By then not only the right side but my whole body felt heavy. Sudden fear rushed through my heart. Did I have a full stroke? Am I going to be bedridden and crippled forever? I had no fear of death. But I couldn't even think of being bed ridden for a few days. From then on I couldn't control my tears.

In the hospital, I was put on oxygen. They gave me intravenous liquids the first day, ran all diagnostic tests on me: MRI, ultrasound, various blood studies, etc. The next day my doctor came and gave me the news that I was afraid of:

"You had a stroke. Two small arteries have been damaged in your brain," he said. "You will be kept in this room until your present condition gets stabilized. Your blood pressure and blood sugar are very high. After you are stabilized you will be sent to a rehabilitation unit where you will be kept until you get well."

"I can't move, doctor; can't you make my hands and leg move?" I tried to say to him but words didn't come out of my throat. The ball - it was stuck in my throat. I felt like choking. Hot tear-beads streamed down both my cheeks. Helplessly, I lay there staring at his face with lips quivering. He couldn't hear anything I was saying. My mind kept reminding me "No use, you lost the war." I felt like I was losing my mind too.

"Please do not cry," he said with concern. "Crying is going to put your progress two steps backward." There was a pause.

"You have to rest now" he added, "Your therapies will begin in a couple of days, then your body will begin to function again."

During the next six days, my blood sugar and blood pressure did not regulate nor reduce from the original reading in spite of many medications. Insulin was administered intramuscularly three times a day before meals with no effect. I continued to get headaches and insomnia. My headache moved from the side to the center of my forehead. The right side of the entire body remained numb and weak. The medication they gave me did not help me sleep.

At that time my daughter (who is a faithful follower of NAET) realized that I was allergic to every hospital food I was eating and the water I was drinking. I wasn't eating much since I couldn't swallow well. I was given jello, puddings, etc. She tested me using NAET testing methods for the hospital foods and other things around me and found me allergic to all those simple foods. She immediately removed the hospital water and food and brought non-allergic water and food from home. She also brought my NAET practitioner to the hospital who gave a treatment for a sample of the combined food I ate in the hospital. She also gave me an acupuncture treatment in the hospital bed behind closed doors three times a day, each session lasting for 45 minutes. As soon as I started NAET treatments for the food and drinks and received a couple of sessions of acupuncture treatments, began using non-allergic foods and drinks, my blood pressure dropped to 160/90 and my blood sugar dropped to 200 mgs/dl. My speech was still slurred. My weakness of the right side continued but I was able to move my right toe slightly. I had no sensation anywhere else on my right side. I was breathing easily and the oxygen therapy was discontinued. On the seventh day, my doctor decided to move me to a rehabilitation unit for a few weeks. But at my family's request I was sent home with the prescription to continue medications, physical therapy, occupational therapy, speech therapy, and 24-hour nursing care.

I was carried to my room. I couldn't move or turn in the bed on my own. My right side was still numb and weak. My speech was still slurred.

Once in the house in my bedroom, my NAET practitioner discovered the cause of my stroke. She found that an allergy to a particular wall-to-wall wooden cabinet (oak) in my bedroom was the cause. It was hard to believe. A wall to wall cabinet that was there in the same room for six years was the cause of a stroke? All my family members except my daughter arched their eyebrows questioning this diagnosis. Energy medicine is still a mystery to my family just like millions in this world doubting its existence and validity. I trusted my daughter. Mu daughter trusted my doctor. She also tested through her special testing and discovered that I had been reacting to this wooden cabinet ever since I moved into the room, but reacted severely for the past three months. This cabinet had the repelling electromagnetic energy of 20 feet distance. On further investigation, it was revealed that this cabinet was installed in my room six years before (about the time my diabetes began). Three months ago I rearranged my bed from a 15 feet distance to closer proximity of 2 feet to the cabinet. By Question Response Testing (QRT), it was also found that I was reacting to the cabinet at different organs, meridians and levels. This cabinet irritated muy spleen and pancreas from 15 feet distance since I moved in and caused my diabetes six years ago. When I moved to a closer proximity of two feet, it began irritating my liver and caused the stroke now. She said that according to Chinese medicine theory, energy disturbance in the spleen causes adult onset diabetes and energy disturbance in the liver causes stroke. Strange it may sound but energy medicine is very different from other medical disciplines.

My doctor also detected that I needed 45 NAET treatments one every hour for the wood to follow by another 45 NAET treatments every ten minutes. She concluded saying, you have a very good chance of complete recovery from this stroke by the time we complete the 90 sessions of NAET.

Tears rolled down my cheeks. I couldn't talk I could only shed tears. I was saying in my mind, thank you doctor, from your mouth to God's hands. But she never heard my words. My words were stuck in my throat unable to come out. NAET treatment for the cabinet was started right away as scheduled.

I continued all the prescription medications. My NAET doctor gave me acupuncture treatment three times a day. Needles were inserted on the right side of the scalp, shoulder, arm, thigh, leg, fingers and toes and all major acupuncture points. At the end of the eighth day, I began having funny sensation (like pins and needles) on my right arm and leg.

By the end of ninth day, after 24 NAET treatments, my fingers and toes began to move and by the evening my hands and legs began moving. My speech cleared. My fasting sugar dropped to 185 mgs/dl; my blood pressure dropped to 150/90 mm hg.

On the tenth day I sat up on my own. I was also able to take one step with assistance. Next day, blood pressure stabilized at about 140/84 mm hg. Fasting blood sugar was 150 mgs/dl. I was able to sit up on my own and eat normally.

It took four days to complete 90 NAET treatments for the wood. By the end of the tenth day 80 % of mobility of my arms, legs and body returned and I was able to walk using a walker in the room.

On eleventh day I was able to walk without assistance inside the house. I was also able to climb a step with assistance. Suddenly I became very emotional. Tears streamed down my cheeks ... this time tears of happiness! Thank you God, You are Great!

I was overjoyed. I knew that I had made a giant leap toward recovery. My children kept me very busy with multitudes of treatments.

I was also given Chinese and nutritional supplements to clean my circulation, along with my prescription medication.

Since I had been treated for the NAET Basic fifteen and many more allergens before, my NAET practitioner could treat the other culprits causing my present health concerns immediately without losing time on the Basic NAET allergens. I had started NAET treatments two years before, hoping to stabilize my blood sugar. Since it didn't make much difference, I stopped getting regular NAET treatments before completing all known allergens a year prior to the stroke. Now I wish I had listened to my practitioner's suggestion to test the whole Basic set of allergens and treat them if necessary. I was never treated for wood mix, that was one of allergens in the Basic set..

I was also given regular physical therapies, speech therapies, and other exercises. My children insisted that I needed everything to get complete recovery.

On the fourteenth day after the stroke I was able to walk on my own to my doctor's office for an appointment without assistance. He almost fell off his chair when he saw me walking into his office unassisted. I was able to sign my name with my right hand; I was able to take care of my daily needs like using the rest room, brushing my teeth, showering with assistance, holding spoons and forks in my right hand etc. I could say that at that point I was 95% recovered. My blood sugar was tested and the test was normal. After six long years my fasting blood sugar was 95 mgs/dl. I took a sigh of relief.

Now when I think back with this new knowledge of NAET, a lot of my unanswered questions got answered.

My NAET practitioner's evaluation of my condition made sense to me. According to her NAET testing, I was reacting to the wood on the cabinet from 20 feet away. My bed was about 15 feet from the wall. According to her, my spleen and pancreas were reacting with that wood from 15 feet away. I was diagnosed diabetic a week after I began living in that room. I continued to live there for six years and I suffered from high levels of glucose in the blood for the last six years. Medications could not reduce the blood sugar because the causative factor was still there near me.

My NAET practitioner solved the mystery for me. She had a different explanation for my strange sudden health problems and now it made complete sense to me. I am so grateful to her for taking time to explain to me about my body, body functions, and the energy pathways in the body–how energy circulation can get affected by the presence of allergens in a person's energy field;

and how these energy interferences contribute to poor energy and sickness. She also taught me how to prevent future occurrences.

As I write this, it is exactly four years and six months after my stroke. Now my recovery is 100%. My fasting blood sugar has remained around 110-120 mgs/dl. My blood pressure is 120/70 mm of hg. My cholesterol is 180 -190 mgs/dl. I sleep well. I don't get any pins and needles sensations. I can cook my meals and do necessary shopping on my own. I still stay in the same room and my bed is still lying close to the cabinet. The wood cabinet hasn't bothered me. I don't wake up with headaches anymore. Generally I am feeling better than before my stroke. Thank You, Devi, Mala, Thank you NAET.

My NAET Practitioner:
Mala Moosad, L.Ac., R.N., N.D.
Buena Park, CA

Sara's case study is an eye opener to all. Not only just to stroke victims, but to anyone with any disease due to unknown cause as in stroke. There is no definite cause for stroke. An allergy may be behind any unsuspected health disorder. Unless NAET is studied more closely, the process is completely understood, and doctors from all disciplines are taught to use it in their daily practice for evaluation and treatment, our present day health problems will be on the rise, by forcing us to spend a large part of our earning on healthcare without receiving the expected benefit.

Let's look at "strokes" alone. No one really knows how many thousands of diagnosed stroke patients are really suffering from "strokes" or the effect of some environmental or chemical sensitivities as in Sara's case.

SOME STATISTICS ABOUT STROKES

Stroke killed 163,538 people in 2001. It's the third largest cause of death, ranking behind "diseases of the heart" and all forms of cancer. Stroke is a leading cause of serious, long-term disability in the United States.

About 4,800,000 stroke survivors are alive today. 2,100,000 are males and 2,700,000 are females.

Data from GCNKSS studies show that about 700,000 people suffer a new or recurrent stroke each year. About 500,000 of these are first attacks and 200,000 are recurrent attacks.

In 2001 females accounted for 61.4 percent of stroke deaths.

Compared with whites, young African Americans have a two-to- threefold greater risk of ischemic stroke, and African-American men and women are more likely to die of stroke. About 47 percent of stroke deaths occur out of hospital.

On average, someone in the United States suffers a stroke every 45 seconds; every 3.1 minutes someone dies of a stroke. About 4.7 million stroke survivors (2.3 million men, 2.4 million women) are alive today.

The estimated age-adjusted prevalence of stroke for Americans aged 20 years and older is 2.2% for non-Hispanic white men and 1.5% for women; for non-Hispanic blacks, 2.5% for men and 3.2% for women; and for Mexican Americans, 2.3% for men and 1.3% for women (NHANES III, 1988-1994). Stroke is the leading cause of serious, long-term disability in the United States.

8% of men and 11% of women will have a stroke within six (6) years after a heart attack. 14% of people who have a stroke or TIA will have another within a year.

22% of men and 25% of women who have an initial stroke die within a year.

Stroke is more common in men than in women. In most age groups, more men than women will have a stroke in a given year.

Sixty-six percent of all strokes occur in people over the age of 65.

From 1991 to 2001 the death rate from stroke declined 3.4 percent, but the actual number of stroke deaths rose 7.7 percent.

The 2001 death rates per 100,000 population for stroke were 56.5 for white males and 85.4 for black males, and 54.5 for white females and 73.7 for black females.

WORLDWIDE STATISTICS

♦ The World Health Organization estimates that in 2001 there were over 20.5 million strokes worldwide. 5.5 million of these were fatal.

♦ High blood pressure contributes to over 12.7 million strokes worldwide.

♦ Europe averages approximately 650,000 stroke deaths each year.

For stroke information, call the American Stroke Association at 1-888-4-STROKE. For information on life after stroke, ask for the Stroke Family Support Network.

WHAT IS A STROKE?

Blood is carried to the brain via a complex network of arteries and vessels. A stroke occurs when one of these arteries becomes blocked or an artery ruptures.

Stroke is a sudden interruption in the blood supply of the brain. Most strokes are caused by an abrupt blockage of arteries leading to the brain (ischemic stroke). Other strokes are caused by bleeding into brain tissue when a blood vessel bursts (hemorrhagic stroke). Because stroke occurs rapidly and requires immediate treatment, stroke is also called a brain attack. Stroke has many consequences. The effects of a stroke depend on which part of the brain is injured, and how severely it is injured. Strokes may cause sudden weakness, loss of sensation, or difficulty with speaking, seeing, or walking. Since different parts of the brain control different areas and functions, it is usually the area immediately surrounding the stroke that is affected. Sometimes people with stroke have a headache, but stroke can also be completely painless. It is very important to recognize the warning signs of stroke and to get immediate medical attention if they occur.

When the symptoms of a stroke last only a short time (less than an hour), this is called a transient ischemic attack (TIA) or mini-stroke.

Poor blood supply to parts of the brain due to various reasons cause temporary disturbances in the brain function. Energy pathways begin and end or go through the brain. Disturbances in the energy pathways cause spasms in the nerve fibers leading to constrictions in the blood vessels. When there are spasms or constriction in any pathway, the function is impaired. Blood supply is limited to the target tissue and the tissue suffers from poor oxygenation. Insufficient blood supply to parts of brain for brief periods causes transient ischemic attacks. In most cases body's defense forces are able to solve the minor unknown causes of TIA without delay and the blood supply is restored quickly and the brain tissue doesn't die or become damaged as it does in stroke. A TIA can last from a few minutes to a few hours; sometimes days. When a TIA progresses more than six days, it is called "stroke."

REDUCING RISK FACTORS

The blood supply is not restored fast enough in stroke and thus causes tissue damage. Repeated TIA can lead to stroke eventually. Strokes are the most common cause of disabling neurologic damage. Some of the known causes of stroke are: high blood pressure, hardening of the arteries from fat build up, high cholesterol, smoking, and diabetes. Food and environmental allergies could be the underlying causes of the risk factors. An allergy to salt and calcium products may be a contributing factor for high blood pressure; an allergy to proteins and minerals may be causing hardening of the blood vessels; malabsorption and allergy to fats can cause fat build up in the arteries; and high cholesterol levels in the blood may be due to an allergy to animal and vegetable fats. An allergy to sugar, insulin, digestive enzymes, and coenzymes can cause poor sugar digestion and utilization can cause diabetes; an allergy to B vitamins and sugars or starches can cause poor digestion, and general fatigue; Poor utilization of B vitamins and sugars can lead to various addictions including smoking, overeating and obesity. We have a few case studies now to show the effect of NAET in reducing hypertension, cholesterol levels, reducing or dissolving arterial plaques, reducing diabetes, helping to stop smoking, helping with elimination of eating disorders, with eating disorders, reducing obesity, and improving overall circulation, etc. We need to do extensive research in these areas to understand the relationship between allergies and reducing risk factors and thus preventing or reducing major health disorders like stroke.

A stroke or TIA can occur when blood flow to the brain is normal but the blood doesn't contain enough oxygen. This can happen when a person gets carbon monoxide poisoning or is sensitive to perfumes, extreme toxic fumes and materials and repeated exposure to toxic substances like insecticides, malathion, and pesticides. One of my chemically sensitive patients complained of internal tremors every time she went to her new job in an office. She also said that she felt numb above her neck and her brain felt like "a piece of lead" by 10:00 a.m. She got off and reached home around 4:30 p.m. By 6:00 p.m., she felt better. She enjoyed her work but this mysterious problem worried her. She always dressed in dry cleaned clothes when she went to work. In her case, the reason for her mysterious problem turned out to be the dry cleaning solvent perchloroethylene. As soon as the allergen was identified, she was treated for the chemical and hasn't experienced the same problem again even though she is still getting her clothes dry-cleaned. Perchloroethylene is a known carcinogen and many people are found to be allergic to this chemical; it is very toxic to many people. The toxic reactions can affect people differently.

SYMPTOMS OF TIA

(Transient Ischemic Attacks)

Loss of or abnormal sensations in an arm, leg or one side of the body

Weakness or paralysis of an arm, leg or one side of the body

Partial loss of vision or hearing

Double vision

Dizziness

Slurred speech

Difficulty in thinking of the appropriate word

Inability to recognize parts of the body

Unusual movements or no movement

Loss of bladder control

Imbalance and falling

Fainting

While the symptoms are similar to those of a stoke, they are temporary and reversible. In the case of strokes, since there is tissue damage, most of the symptoms are difficult to reverse.

OTHER REACTIONS TO ENVIRONMENTAL ALLERGENS

Some people with allergies experience symptoms close to TIA. If the allergies are left untreated they can cause repeated TIA-like symptoms due to poor circulation to the affected tissue. If TIA-like symptoms are repeated they can cause progressive weakness and damage to the tissue and eventually a TIA can have a deeper effect leading to stroke.

We know that symptoms like these experienced by people with environmental allergies. Whenever this group of people come near the electromagnetic energy field of a particular allergen, their nervous system experiences nerve energy disturbances throughout the spinal cord and the energy pathways. But stroke-like symptoms are only part of what they might experience. Many people complain that they have internal tremors on and off and some people suffer from the tremors of the whole body. Some others have reported that they feel their spine trembling most of the time. If the allergen affected the liver and gall bladder meridian (Oriental medical theory, please read Chapter 5 for more explanation on this subject) and caused energy disturbances in those meridians, the patient would feel tremors in the body areas depending on the affected meridian(s). If the brain is affected, the patient would feel energy disturbances in the brain. Food allergens most often affect the body organs first (stomach, spleen and colon meridians). Unlike food allergens, the environmental allergens (like wood, smoke, latex, perfume, pesticides, silica, grasses, weeds, etc.), affect the brain and nervous system first (liver, gall bladder and kidney meridians).

This is the reason why the environmentally sensitive people, whenever they are exposed to toxic environments can experience frequent brain symptoms like brain fog, poor clarity of thinking, difficulty in thinking of the appropriate word or saying it, fainting, double vision, disturbed vision and hearing, imbalance and falling, poor memory, dizziness, and abnormal sensations in different parts of the body. Their level of well-being changes from minute to minute. They may feel miserable at one point (when they get closer to the allergen(s), then suddenly they may feel somewhat better for a while (whenever they move away from the presence of the allergen(s). Many people live their everyday life going through these "feel-bad feel-good feel-fair feel-miserable" cycles, one may call it "roller-coaster" like life. The intensity of the symptoms may vary in people depending on the affected energy pathways, the area and amount of tissue involvement, age, immune system status, and the duration of the problem.

An allergen's energy initiates an energy disturbance in the weakest part of the tissue in the body (a tissue: muscle, bone, or an organ, etc.). The affected tissue is connected with one of the major twelve meridians (Read chapter 5 to learn about meridians). The affected tissue will try to fight the harmful energy by encouraging the production of various chemical messengers like antihistamines, corticosteroids, prostaglandin, adrenaline, cytokinins, endorphins, enkephalin, heparin, neurokinin, and interferon, etc., whichever are appropriate to the condition to counteract the adverse reaction from the invading energy of the allergen. If the body can create all these immune system modulators on the brain's command in a few seconds, then the body will come to a settlement with the allergen in seconds without causing any obvious ill-health symptoms in the person. Unfortunately, this only happens in a person with a normal immune system. But when the immune system perceives what should be a harmless substances as a dangerous intruder and stimulates antibody production to defend the body, then things do not settle down as pleasantly.

Here, the first contact with an allergen initiates the baby step of an allergic reaction. The body will alert its defense forces in response to the new invader and will immediately produce a few antibodies, storing them in reserve for future use.

In most people, a first contact or initial sensitization will usually not produce many symptoms. During the second exposure to the allergen, the body will alert the previously produced antibodies to action, producing more noticeable symptoms. In a person with an average immune system, the second exposure may not cause too many unpleasant symptoms, either. But often with the third exposure, the threatened immune system will begin serious action causing allergic reactions. During this process various types of antibodies are produced.

IgE - is the antibody that is responsible for immediate hypersensitivity and reactions in the body when it comes in contact with an allergen.

Some people could have IgE mediated allergies only, in which case IgE antibodies can be found in the blood samples whenever their blood serum is tested by RAST. This also indicates that these people are having repeated severe reactions on a daily basis on a variety of items. When a person with previously tested high IgE level receives a lower or normal reading on the second testing, that is an indication for many things:

The patient may be allergic to one or two items only and chances of exposure to them are minimal. The patient may be avoiding the known allergens completely (recent exposure is needed for IgE production).

The patient may have had an allergy desensitization treatment like NAET between the two tests.

An allergic reaction may be manifested in varying degrees as mild to severe itching, rashes, hives or swelling anywhere within the body, restricted blood circulation in certain areas leading to heart attacks and even stroke.

Now let us evaluate using the meridian system from Oriental medical principles: The affected tissue will prevent the entry of the adverse energy by constricting the entrance of the meridian. This constriction may feel like a pain, spasm, or some form of discomfort. If the tissue wins the fight with the help of defense forces, the adverse energy will leave the affected tissue without any of the above mentioned reactions. If the affected tissue failed, the invading energy will enter the affected tissue's meridian first, then it will expand into its paired meridian; and continue to invade the neighbor meridians until it takes over all 12 meridians which cover the entire body. Thus the imbalance will initiate in the affected tissue, then the energy disturbance will spread through the meridians causing energetic imbalances thus causing allergic reactions. These energetic imbalances are called diseases. If one could locate the unsuitable energy (allergen) that initiated the cycle of energy disturbance, if that allergen is deactivated or desensitized through NAET, all other meridians will clear their energy disturbances instantly as though a switch is thrown and the body returns to it's original state. It is an art to locate and identify the initial causative agent. Understanding the categories of allergens, the effect of the allergens, and associated meridians, help us identify the causative allergens.

Everyone's body has a 'Turn on' and 'Turn off' mechanism. Certain items at certain times turn on certain functions in the body. In the same way, certain items turn off certain functions in the body. No one knows exactly what turns on and what turns off or when and why it happens.

Allergens can greatly influence the "Turn on" and "Turn off" mechanism. This can affect various parts of the body and cause energy disturbance in the sufferer. The manifesting symptoms will depend upon the area affected by the presence of the allergen in one's electromagnetic field. An allergic reaction can take place in four different ways:

1). when one's body is within a few feet of the electromagnetic field of the allergen (as in touching, standing, or sitting near the allergen);

2). by direct contact when one is contacting the allergen physically (by wearing clothes, jewelry, cosmetics, etc.);

3). by taking it inside the body as in eating food, drinks, medicine, herbs, vitamins, etc.

4). as an emotional response from an interaction of the body with the substance as in thinking negative thoughts or saying negative remarks or experiencing negative interactions..

When the allergen's energy tries to pass through one's energy pathways, if it faces a resistance of some sort in the pathway of the energy flow (resistance may be due to an unwelcome gesture from the nearby tissue produced as a muscle spasm, a kink in the nerve fiber, a scar from an old wound, surgery, etc.), its flow gets disturbed or stopped by the resistance, then it will try to find an alternate route if there is an alternate open, or reverse the flow in case a complete blockage is found ahead. If there are more blockages in front and behind; if it cannot flow in any direction, it stops. Along with the energy flow, the allergen's energy will also settle in the area where the energy circulation stopped. When the energy circulation stops, many associated functions of that energy meridian will stop also. When an area of the body struggles to receive appropriate nerve energy supply for a certain period of time, it also struggles for oxygen, nutrition, and communication with the rest of the parts of the body causing tissue damage and diminished or abnormal function of the area, which we recognize as various present-day health disorders. According to Oriental medical principles, this is the way a foreign energy initiates energy flow interference leading to minor to severe ailments and disorders..

Let us view or evaluate this theory and understand it using our Western medical mind.

At the present time, Western medicine mostly recognizes the symptoms, pains and discomforts that have direct relation with the affected tissue. In certain situations some referred pains are taken into consideration as in a case of gall stone: pain referred to the mid-thoracic area; or in a case of sciatic neuralgia: pain radiating to the knee or ankle, etc. because, Western medicine views everything according to the dermatomal distribution or autonomic nerve distribution and these are visible nerve pathways. Western medicine hasn't been able to scientifically prove the existence of meridian theory yet, even though a few studies have been done using Kirlian photography which established the existence of the meridian system.

Whether we have proved the meridian theory or not, we all know that vital energy flows through our body when we are alive, and it stops flowing when we die. No one has yet been able to map or scan the travel route of that electrical energy system before and after death and package it to present it to the world. But the Chinese pioneer researchers have done this by scanning the body energetically, studying the meridian system, and understanding the relationship between the normal energetic system with that of sick people. To understand the associated tissue of the body in relation to pain or blockage, one needs to understand the meridian system. If you are interested to expand your understanding, please read any book on acupuncture that is listed in the bibliography.

Any blockage in the energy pathway, organ, or vessel is a physical blockage that is capable of reducing the energy flow. For example: a gallstone in the gallbladder, or a uterine fibroid, a plaque in a blood vessel, a surgical scar, acne, growth, tumor anywhere in the body, an abscess, wound, cut, even a boil on the skin surface is capable of causing an obstruction in the energy flow. We can all visualize that.

GALLSTONE AFFECTING THE ENERGY CIRCULATION

Let us visualize the energy pathway like a small stream with crystal-clear-water flowing. A blockage (a gallstone in the gall bladder means a blockage is in the gall bladder meridian) can be like a rock lying in the stream causing an obstruction for the flowing water. When the water flows past the stone it has to circle the stone to flow away because the solid stone won't move out of its course. If the stone is a little one, flow of the water doesn't get disturbed since it can flow on top of it or will encircle it without much effort. But if it is huge, the flow gets disturbed and will find an alternate route or it will stop flowing, creating erosion or water damage in the nearby areas. Similar things take place in our body. A physical blockage is not going to move out of the energy route if it is composed of solid tissue (e.g. a gallstone). It could cause the body to experience pain and discomfort when the energy tries to force itself through the blockage. Let us see how a gallstone can cause physical, physiological and emotional disturbances in the body: impaired physical function due to a gallstone in the gall bladder could cause pain on the right side of the abdomen, because a physical blockage can give physical symptoms as localized pain anywhere in the travel route; when the stone is left unattended for months or years, it can affect the physiological function causing indigestion, heart burn, nausea, etc., poor digestion and assimilation of the vital nutrients; when the body is not receiving vital nutrients, its enzymatic functions suffer and will begin producing abnormal enzymes. The abnormal enzymes produce abnormal reactions in the body and nervous system functions causing disturbances in the cellular levels (or we may call this psychological, emotional, mental, or spiritual). Some of these cellular level disturbances are: depression, anger, mood swings, feeling of insecurity, frequent crying spells, paranoia, personality disorders, etc. Whenever the energy is stopped from flowing towards the target tissue or organ, the tissue will not get nerve supply or appropriate nutrients; with the result, physiological and psychological disturbances will develop. A physiological response can cause poor tissue function at any of these three levels (physical, physiological or psychological) anywhere in the travel pathway.

If we could scan the body, or take ultrasound, MRI or CAT scan, X-Rays, etc., we could see the evidence of blockage because of a gallstone. Since this solid extra substance is lying in the energy pathway, the energy has to be rerouted until the obstruction is removed. The area with an obstruction will be the weakest area of the body.

A-71- year old female patient came in with the complaint of occasional muscle pulling pain in her right big toe and fourth toe. She had no other symptoms. A body scan would find the presence of a gallstone in the gallbladder. When one understands alternate medical principles, the reason for pain in the big toe and fourth toe would be understood to be gallstone in the gallbladder, obstructing the energy flow in the gallbladder and liver meridians.

How does this knowledge help NAET work with allergies? AS stated, if there is a physical blockage, when the energy flows through that area, it creates energy turbulences, which we can detect without complicated scans. In this case, when the allergen's energy flows through the liver and gallbladder meridians, symptoms of energy disturbance in those meridians have been manifested in the patient, which we can track down as "a weak NST" (read Chapter 6) on the liver and gallbladder meridians. We call this weak NST as "an allergy." As I have stated before, allergy effects the weakest tissue of the body. Because of the presence of the immovable gallstone, this particular meridian is the weakest meridian in this body. Liver is the sister organ. So whatever happens to the gall bladder meridian is being shared by the liver meridian and the physical pain is felt in both meridians.

UNDERSTANDING ALLERGIES

We know about certain risk factors that can cause as example, heart attacks, but some people may not be affected from those risk factors at all. For example: cigarette smoking. Lets examine two groups of people:

Group A may have spent their entire life as chain smokers yet never get any heart problems, cancer or any other diseases associated with smoking.

Group B may have only smoked for a few years, just a few cigarettes a day, yet they may suffer from lung cancer or heart attacks or strokes. When we test this two groups using NAET testing methods, we will find out why these two groups respond differently:

Group A did not have an allergy to cigarettes whereas group B was severely allergic to them. The allergy to cigarettes made their organs weak, meridians weak, the target tissue weak, and the immune system weak causing diseases starting with the weakest tissue or areas of the body. If the heart was the weakest area in a person, eventually after long-term use, this person could get heart disease complications (high blood pressure, heart attacks, etc.); if the lungs were the weak ones, then he/she could get bronchitis, pneumonia, cough, or cancer; if the spleen was the weak one, then he/she can become obese, diabetic, hypoglycemic, or can get blood vessel disorders or bleeding disorders.

If done properly, NAET testing can give specific answers to puzzling problems. NTP to detect allergies are described in Chapter six.

A-48- year old female patient, Donna, was suffering from diabetic neuropathy for two years. Her complaints included: poor sensation in her feet and hands especially on the tip of her toes and fingers (stocking-glove distribution), numbness, tingling and paresthesia (pins and needles-like sensation) of the extremities, occasional right sided sciatic neuralgia (radiation of pain from right side of the hip to the ankle joint), frequent ankle jerks on both sides but more pronounced on her right leg, fatigue after meals, and insomnia. When she was evaluated in our office, we detected an allergy to the prescription medication she was taking twice a day to control her diabetes. She was very uncomfortable with her pain and other symptoms. We treated as an acute case for the medication right away before clearing the NAET Basics. After 36 hours (that was the time needed for her to overcome the allergic reaction to that particular drug, according to our test), she returned to the office with a big smile on her face. Her diabetic neuropathy of two years completely subsided four hours after NAET. She then went through the complete NAET program. Three years later after completion of NAET treatments, her blood sugar maintains in a high normal range (fasting:100-110 mgs/dl) with diet, exercise, eating and using allergy-free substances and no trace of previous symptoms.

Allergy has probably been in existence from the time the world began. Early on, though, unusual reactions were not referred to as "allergies." No one connected the underlying causes of these numerous health disorders with the food we eat daily or the substances we use or come in contact with in our everyday life. Everyone is unique and everyone reacts differently and has different symptoms of allergy. This makes it difficult to see the connection between the substances that we use and the health disorders we suffer. So, instead of investigating the reactions to our living environments, medical professionals began calling these strange reactions by different medical names.

Somewhere around mid 400 BC, Hippocrates, the Greek physician considered the "Father of Medicine," noted that cheese caused severe reactions in some men, while others could eat and enjoy it with no unpleasant aftereffects. Three hundred years later, the Roman philosopher Lucretius said, "What is food for some may be fierce poison for others." From this we derive the expression, "One man's meat is another man's poison." This is a simple and concise definition of allergy, although the condition was not recognized as "allergy" until recently.

An allergy is a hereditary condition: an allergic predisposition or tendency is inherited but may not be manifested until some later date. Researchers have found that when both parents were or are allergy-sensitive, 75 to 100 percent of their offspring react to those same allergens. When neither of the parents is or was sensitive to allergens, a probability of producing allergic offspring drops dramatically to less than 10 percent.

Inherited nerve energy supply from certain sensory nerve receptors to certain organs often remains dormant and unable to conduct messages to and from the spinal cord and the brain. In some instances, these dormant receptors become hyposensitive toward certain items, whereas in other people they become hypersensitive. Neither type functions normally.

It has also been proven that the age of onset of an allergic condition definitely depends on the degree of inheritance. Thus, the stronger the inheritance, the earlier the onset.

Major illnesses, severe reactions to drugs, toxins, chemicals, radiation, overexposure to toxins, toxic chemicals, severe emotional traumas, etc., are capable of causing damage to the sensory nerve fibers and marring their conductivity.

It is estimated that 90 percent of the population throughout the world suffer from allergies. But the estimate is just that, an estimate, due to the various definitions of allergies among researchers. If medical researchers were willing to broaden their views on allergies to include hypersensitivity, intolerances, IgE mediated and non-IgE mediated reactions, we would clearly recognize the overwhelming percentage of allergy sufferers.

The shocking fact is that there are hardly any human diseases or conditions in which allergic factors are not involved directly or indirectly. Any substance under the sun, including sunlight itself, can cause an allergic reaction in any individual. In other words, potentially, you can be allergic to anything you come in contact with. If you begin to check people around you—even so called healthy people—you will find hidden allergies as a causative factor in almost all health disorders. You saw earlier in this chapter (case history); how a stroke was traced back to a simple allergy. How could we prove the stroke was in fact caused from an allergy? Proof is in the results.

We cannot ignore the fact that we are in the twenty-first century where technology is even more predominant than ever. There is nothing wrong with technology; modern technology has provided us better quality of life. But the allergic patient must find ways to overcome adverse reactions to chemicals and other allergens produced by the technology, in order to live a better life in this new world. Even though it requires a series of detailed treatments, NAET offers the prospect of relief to those who suffer from constant allergic reactions by reprogramming the brain to perfect health. Just like resetting a computer, we can reset our nervous system through NAET and overcome the adverse reactions of the brain and body to the technological advances (chemicals, plastics, electronics, etc.) and also to food, cosmetics, fabrics, animals, etc.

Illness is a warning given by the brain to the rest of the body regarding energy blockages within the energy channels. Through illness, pain, inflammation, fever, heart attacks, strokes, abnormal growths, tumors and various discomforts, the brain signals the body about the possible dangers if the energy blockages are allowed to continue within the channels. If the symptoms are minor, blockages are minor, or we can say, if the blockages are minor, symptoms are minor. If the symptoms are major, the blockages are major. Minor blockages can be unblocked easily, whereas major blockages take time and a series of NAET treatments.

Through 31 pairs of spinal nerves, the brain operates the largest network of communication within the body. Energy blockages happen due to contact with adverse energy of other substances. When two adverse energies come closer, repulsion takes place. When two similar energies get together, attraction takes place. The result of the repulsion of energies is referred to as allergy in this book.

I developed this new allergy elimination technique, employing the knowledge from different fields of medicine, to identify and treat for the reactions to many substances, including food, chemicals, and environmental allergens.

Through many long years of research through Western medicine, chiropractic, acupuncture/acupressure, Oriental medical knowledge, nutrition, and after many trials and errors, I devised this techniques to eliminate energy blockages (allergies) permanently and to restore the body to a healthy state. These energy blockage elimination techniques together are called Nambudripad's Allergy Elimination Techniques or NAET for short.

We need the complete cooperation of the whole brain and nervous system to get the best results with NAET, because NAET involves the whole brain and its network of nerves, as it reprograms the brain by erasing previously harmful memory regarding an allergen(s) and imprints the new, useful memory in its place.

When people are allergic to nutrients and foods they are often found to be allergic to environmental factors too. The reason for this is quite simple. Most of our nutrients come from environmental substances. For example: vitamin C is seen in all living plants, trees, flowers, grasses, pollens, fruits, vegetables, etc. So, for example: If one is allergic to vitamin C, (one of the best sources of antioxidants, and a very important nutrient in our daily foods), he/she might be allergic to grasses, weeds, trees, flowers, etc., since vitamin C is highly concentrated in all these. When he/she is treated for NAET sample vitamin C mix, most people who were allergic to grasses, weeds, flowers, trees, etc., will find their allergies to them greatly decreased. When patients are treated for egg mix, calcium mix and mineral mix many clear their sensitivities to pollens, showing that proteins and minerals are concentrated in pollens. When the patient is treated for B vitamins, his/her reactions are reduced towards grasses, since B vitamins are concentrated in grains, and the grains belong to the family of grasses. When the patient is treated for yeast mix, his/her reactions are reduced towards candida, molds and fungus. When patients are treated for all the NAET Basic samples, they begin to absorb normal essential nutrients and improve their immune system.

NAET BASIC ESSENTIAL NUTRIENTS

Egg Mix (animal protein, body protein)

Calcium Mix (milk and milk products and other calciums)

Vitamin C Mix (fruits, vegetables, and other vitamin C products)

B Vitamins (15 B vitamins, bran from whole grains, and wheat)

Sugar Mix (16 different sugars: natural sugar, body sugar like glucose and dextrose, etc.)

Iron Mix (dark green vegetables, dates, beef, red meat, etc.)

Vitamin A Mix (beta carotene, fish, shellfish, etc.)

Mineral Mix (10 macro minerals and 44 trace minerals)

Salt Mix (table salt, sodium chloride, etc.)

Grain Mix (various whole grains)

Yeast Mix (different types of yeast including baker's, brewer's, and candida albicans)

Stomach Acids (highly acidic stomach secretions)

Digestive enzymes or Base (intestinal digestive enzymes)

Hormones (female and male hormones)

Fatty Acids (essential fatty acids)

If you can digest, assimilate, absorb or correct with NAET these above 15 groups you will have a good immune system.

When you have a good (normal) immune system, your bodily functions take place normally without throwing the body into hyperactive or hypoactive state. The body experiences a state of homeostasis and all the cells, tissues, organs, nerves and the network of nervous system function in a synchronized, unified fashion where the entire body as a unit enjoys the bliss of living. It is very important to treat the NAET Basics before treating for environmental allergens. When patients eliminate any allergies to NAET Basic essential nutrients, their reactions towards the environment reduce in many fold. We NAET professionals can then tackle whatever environmental allergies remain, knowing that the core elements that impact many individuals are out of the picture.

2

Categories of Allergens

People who are sensitive to environmental allergens can be divided into two groups: The first group can suffer from allergies to any natural environmental substance like pollen, grasses, weeds, wood, trees, flowers, mold, animal dander, animal epithelial, animal droppings, human hair, sand, precious stones, dirt, dust, insects, wool, cotton, silica, salt, wholesome nutrients, rain-water, weather change, heat, humidity, dampness, cold, wind, high altitude, low altitude, ocean water, etc. and all other natural creations of the world. Some even react to foods prepared from natural sources. They cannot eat vegetables, whole grain products, unprocessed nuts, etc., without having severe reactions. They react less to processed foods and to foods with less natural nutrients. They cannot eat organic foods without getting sick.

The second group suffers reactions from man-made chemicals and other chemicals seen or used on the environment. For example: pesticides, weed killers, fertilizers, insecticides, and plant food, cleaning chemicals, etc. This group is able to eat natural foods (without any chemicals on them) without any adverse reaction. But if the food is prepared in plastic containers, if the food is prepared with unnatural items like food colors, food additives, food preservatives, or any other chemicals, they might have allergic reactions. As long as they wear or use 100% cotton, silk or any natural fabrics, they function well. If they use any synthetic materials like polyester, acrylic or acetate, they become dysfunctional. We can call this group "chemically sensitive" people. Both groups face nemerous hardships to live in our present world. We will be discussing mostly about the first group in this book. The second group will be discussed in the book, *Freedom From Chemical Sensitivities.*

Then there is a third group who reacts to everything around them. Environments as well as chemicals from the world make them sick. These are universal reactors.

Commonly seen allergens are generally classified into nine basic categories based primarily on the method in which they are contacted, rather than the symptoms they produce.

1. Inhalants

2. Ingestants

3. Contactants

4. Injectants

5. Infectants

6. Physical agents

7. Genetic factors

8. Molds and fungi

9. Emotional Stressors

INHALANTS

Inhalants are those allergens that are contacted through the nose, throat and bronchial tubes. Examples of inhalants are microscopic spores of certain grasses, flowers, pollens, powders, smoke, cosmetics, dust, dust mites, flour from grains, perfumes; fumes such as smoke from fireplace, burning wood, pollution in the air; smell from flowers, smell from cooking, fermented foods, body odor, body secretions of self or others, etc.

Usually there is no predictable allergic reaction, or set of reactions, in response to a given allergen. A sensitive person can react to an allergen adversely at any time. If there is a predictable response, however, it is in this general category of inhalants that it comes closest to being found.

Most of us have suffered the discomfort coming from smelling bad things like decayed substance, stale food, or smell from animal droppings when visiting an animal farm. Most people with normal reactions may not enjoy the smell but when they leave the area, there won't be any further problems. In environmentally sick people, the smell arising from a dirty gutter or animal farm or from a flower bed can give rise to a variety of physical, physiological or emotional symptoms with varying intensities. Sometimes, these smells can produce watery eyes, runny nose, coughing, bronchial spasms, asthma, fainting episodes, or even anaphylaxis. The smell receptors have become the weakest tissue in this group and they have become supersensitive to natural smells from the environment.

Hay-fever is generally the result of breathing the spores of pollinating grass and weeds. It normally occurs when these plants are in bloom in the spring or in warmer climates, closely following a summer rain and the resultant regrowth of the grasses and weeds on the hillsides. Sinus drainage and restricted breathing are the direct and reproducible results of an allergic reaction to an inhalant.

Dust and dust mites cause not only respiratory disorders but are seen in health problems involving other organs and tissues. Carpets, bed, bed linen, pillows, fabric covered sofas and other furniture, draparies, car seats, etc. cause severe health problems in sensitive people. Commonly seen health problems related to respiratory system arising from an allergy to dust or dust mites are other than hay-fever, sinusitis, bronchitis, pneumonia, postnasal drip, and chronic cough. Other symptoms include: arthritis especially the knees joints, itching, hives, headaches while lying in the bed or using pillows, insomnia, pin prick sensation anywhere in the body, etc.

Carpets, pillows and beds accumulate millions of dustmites. Energy fields of sensitive people do not like to share bed and pillows with these microscopic invaders. Pillows and bed linens are easy to wash frequently to kill the dust mites and their eggs. Beds are not easy to wash frequently. Beds can be completely covered with thick plastics. Then the beds should be protected with fully covered mattress covers and bed sheets. These can be washed regularly. Air purifiers and room purifiers or humidifiers can help clean the bedrooms. If possible rooms should be aired out by opening the curtains and windows. Sunlight can help kill dust mites.

One of my patients complained of headaches that was quite unusual. She was woken up with pounding headaches around 3:00 am every morning since four years. First two years she suffered the headaches. Then she decided to see a doctor. After trying different treatments for headaches and without getting any results for next two more years she was referred to us by one of her friends. Her history pointed towards something in the bed causing the headache. She was treated for the bed linen, pillow cases, etc.

through NAET. But she continued to get headaches. The she was asked to bring her pillow. She was allergic to her pillow. After the treatment for her pillows, her headaches reduced to once a week or so. She had various other allergies. So she continued with the NAET basic treatments. After six months, she began getting headaches daily again. By the time we had completed all NAET Basic essentials. We tested her for dust and dust mites and found her to be very allergic to them. After treating for dust and dust mites, her headaches subsided completely.

Another patient suffered from chronic cough for over four years. She was allergic to her carpet. She said the carpet was about ten years old. She did not have the cough for six years. She could not understand why her cough started only after six years. She wasn't convinced that carpet could be the cause for her cough. So, we did not treat her for the carpet. Instead, she was advised to replace the old carpet. We wanted to show her how much she was affected by the carpet and just by avoiding the suspected allergen how much improvement she could receive. She decided to do that before receiving any NAET for the carpet. When the carpet layers came in and removed the carpet she said there were millions of black eggs and tiny worms crawling under the carpet. She decided to put ceramic tiles in the bedroom instead of the carpet. A week after replacing the carpet she returned to the office. She was completely free from her cough for the whole week. But when she returned to work, her cough returned. She had carpets in her office. Since the allergy to carpet and carpet mites were left untreated, eventhough she was not coughing in her house she was bothered by rest of the places with carpets. She couldn't ask everyone to replace their carpets with tiles. So this time she was desperate and wanted to be treated for the carpet allergy. Of course, she had to be treated for the NAET basic essentials first before she was treated for the carpet, mites, dust and dust mites. After successful completion of these treatments she said none of them including new or old carpets, dust or dust mites, or dusty areas make her cough anymore. But she said that she makes sure she wears slippers while walking on carpets just in case some of the worms from beneath the carpets decided to crawl into her clothes she wouldn't know. She didn't want to take any more chances.

Environmentally sensitive people have delicate emotions. Any simple emotional incident can sometimes imprint a lasting image in their minds causing myraid health problems later. I have treated hundreds of environmentally ill people and 90 percent of them had emotionally traumatic past. In most cases, some traumatic emotional event from the past may have triggered the switch to turn on to react to good and bad environments equally. Probably the brain used this defense mechanism in order to protect the body, but when the need completed, brain could not turn it back to original state. Through NAET when we address the exact issue from the past that triggered the signal to react adversely and help balance the emotional status, they would stop their exagerrated reactions to the environment and they will be able to resume normal life.

Environmental illnesses are often misunderstood by families, friends and even by physicians. Most get diagnosed as psychiatric cases, and treated as such with antidepressants, mood elevators or sedatives to assist with symptoms. Few takes time to investigate the actual problems.

A young wife moved to Los Angeles from a rural village in India. She and her husband lived in a small apartment. For the first time in her life she started using natural gas for cooking. The house was also heated by gas. In a few days she became very tired, depressed and began to sleep almost the entire day. Lacking energy or enthusiasm to do anything, she was unable to handle her daily chores. Worried, her husband took her to different doctors. She was treated for psychological problems and placed on antidepressants, which gave her some relief. Then she started having severe nasal congestion, headaches and backaches. At this point, she visited our office. After elimination of a few allergies by NAET, her problems were pinpointed to natural gas. Her system was getting poisoned with natural gas. After she was treated for natural gas by NAET, she improved both mentally and physically. She did not need antidepressants any more.

Environmentally ill people may have many emotional issues along with their physical and physiological problems. Environmental sensitivities make their emotional aspects very weak. It is highly recommended to refer them to professionally trained mental health workers to receive emotional support therapies and counseling along with NAET treatments. They need strong support either from family or friends to help them emotionally while they go through NAET desensitization process. NAET specialists are usually very busy and they may not have enough time to counsel these patients. Patients can feel insecure, paranoid, suspicious of everything around or everybody around them. They may need a close friend to discuss their fears. If they have good family support, spousal support, they will get better soon. They could also join a support group. This is the reason I suggest that they join a support group, yoga group, or see a therapist regularly until they complete NAET treatments for all known allergens.

INGESTANTS

Ingestants are allergens that are contacted in the normal course of eating a meal or that enter the system in other ways by mouth and find their way into the gastrointestinal tract. These include: foods, drinks, condiments, drugs, beverages, chewing gums, vitamin supplements, etc. We must not ignore the potential reactions to things that may be touched and inadvertently transmitted into the mouth by our hands.

The area of ingested allergens is one of the hardest to diagnose, because the allergic responses are often delayed, from several minutes to several days, making the direct association between cause and effect very difficult. This is not to say that an immediate response is not possible. Some people can react violently in seconds after they consume the allergens. In extreme cases, one has to only touch or come near the allergen to signal the central nervous system that it is about to be poisoned, resulting in symptoms that are peculiar to that particular patient. Usually, more violent reactions are observed in ingested allergens than in other forms. Processed or unprocessed tree nuts, eggs, milk, fish, herbs, herbal teas, all kind of spices, etc. can cause immediate or delayed reactions, giving rise to various health problems.

Salina suffered from severe sciatic pains for the last six months. She had various medical treatments before she found us through the web site. On her first visit, NTP determined the cause of her six-month-old sciatic pain to be special expensive herbal tea she had started drinking regularly six months before. She was advised to avoid the tea for the time being and bring the tea to the office on the following visit. A week later when she returned for the second appointment, she reported that by avoiding the tea alone for a week she felt about 80 percent relief from her sciatic pain. She was completely relieved of her pain after successful NAET treatment for the tea.

CONTACTANTS

Contactants produce their effect by direct contact with the skin. This is a problematic area for all environmentally sensitive people. It is not easy to sit at home all the time. People with environmental problems find it hard to take a walk in the park or woods without getting shortness of breath or without triggering an asthmatic attack. They also often suffer depression, panic attacks, nervousness, queasy stomach, unexplained fear, feelings of insecurity, helplessness, diarrhea, abdominal cramps, internal tremors, insomnia, fibromyalgia, general fatigue, poor appetite, weight loss, to name a few. The commonly reactive environmental contactants from natural environments include the well-known poison oak, poison ivy, poison sumac, cats, dogs, also rabbits, wood furniture, wood cabinets, ornamental iron works, iron skillets, utensils, trees, grasses, weeds, pollens, flowers, smell of flowers, early morning dew, natural fabrics like cotton, silk, wool, jute, feathers, herbs, various types of plant oils, petroleum jelly, lanolin hand cremes, ceramics, dirt from the yards, insects, animal dander, epithelial, animal droppings, human hair, rocks, stones, precious stones (diamond, emerald, crystal, etc.), spring water, drinking water, silica, silicone, glasses, sand on the beach, ocean water, ocean air, crude oil, products made from crude oil, smog, ozone, carbon dioxide, carbon monoxide, air pollution, radiation, heat, cold, humidity, dampness, sunlight, etc.

Elsie, 28, suffered from severe fibromyalgia ever since she was a child. Her family was very health conscious. They ate organically grown vegetables, fruits, whole grains, unprocessed nuts and seeds, and dried beans. They were vegetarians. They also wore cotton, silk and wool. The carpet was made from 100 percent cotton, and so were curtains and upholsteries. When Elsie began having strange general body aches around the clock, her parents became more strict in adhering to the organic produce and products. But she lived in misery. Nobody knew the cause of her mysterious illness. Some doctors thought it was genetically transmitted, but none of her immediate ancestors had any such problems. Their family priest thought her problem was perhaps from evil spirits. To give the benefit of the doubt, her family even put her through the ordeal of exorcism. But she continued with her nagging 24-hours fibromyalgia that did not respond to any pain medications or other remedies. Eventually, she accepted her pain as part of her life.

Since high school, she worked with a small sewing company with all natural fibers like cotton, wool, silk, etc. When the company moved to Anaheim, California, she moved with it.

The company owner suffered from asthma and was advised by one of his friends to try NAET. His history revealed that he had suffered from asthma since he bought this sewing company 19 years before. He was on asthma medications with minimal relief. His asthma was worse at work. We found him allergic to all fabrics and materials at work. We began NAET on him. Slowly he began responding. Whenever he was treated for work materials he had to stay away from the company for the following 25 hours. Elsie watched his progress and decided to try NAET on her mysterious illness.

She was found to be highly allergic to all essential nutrients like vitamin C, B vitamins, etc. She was also allergic to all natural fibers like cotton, silk, wool, etc. NAET can only treat one group of allergen at a time. Since she was highly allergic to the items, each item took multiple office visits to get completely desensitized. We suggested that she eat simple foods without packed nutrients, like polished white rice and white flour pastas, etc. until she was treated because she was highly allergic to nutrients. She could not digest brown rice, whole wheat, raw vegetables, raw seeds, nuts, dried beans, etc. She did not produce the digestive enzymes to digest these wholesome complex nutrients in her body. Lack of digestion of these nutrients was the cause of her fibromyalgia. Since she was very allergic to amino acids and other essential nutrients, enzyme supplementation would make her fibromyalgia worse. Such people will do better with predigested food or food without much nutritional value. We also advised her to wear clothes made from polyester since she did not show an allergy when tested by NTP. She was a good follower. She changed her wardrobe from all natural fabrics to synthetic fabrics, ate simple foods and within a week felt 80 percent relief from her nagging constant body aches. She continued to work at the same place but she still felt bad while at the job. Since she knew she was allergic to certain fabrics, she avoided them at work as much as possible. Eventually she was successfully treated for all nutrients, natural foods, and natural fabrics. She became free from her lifelong fibromyalgia. She continued to work with the same company and became free to eat anything she liked without having to worry about the pain.

Julia, 26, was found to have had multiple sclerosis for the previous five years. When she came to our office, she was unable to walk without assistance, and she was almost blind in both eyes. In her case, her silicone breast implant was the cause. When she was cleared for the silicone implant, her symptoms got better. She regained her sight, she became steady on her feet, was able to pass the driver's license test and drive again.

Allergic reactions to contactants can be different in each person, and may include: angina-like pains, anxiety, arthritis on the wrist and finger joints, asthma, carpal tunnel syndrome, chest pains, chills, constipation, cough, cramps in the legs, cramps in the lower abdomen, crying spells, depression, dizziness, eczema, fainting spells, fibrocystic breast, frequent urination, hay-fever, headaches, hives, infertility, insomnia, itching, joint pains, lower back ache, mental confusion, mental fog, mental irritability, mid backache, migraine headaches, muscle spasms, numbness, pain anywhere in the body, pain in the cheek, pain in the hip, pain in the knee, pain in the ankles, pain in the calf muscles, paranoia, pin prick pains, prostatitis, sciatic neuralgia, shivering, sinusitis, skin cancer, skin rashes, sneezing attacks, pain at the sole of the foot, stomach aches, swelling of the body, tennis elbow, uterine fibroids, various kinds of arthritis, vulvodynea, etc. to name a few.

It is apparent that something contacted by the skin can produce symptoms as devastating to the patient as anything ingested or inhaled.

I have treated many patients with above listed health problems. Even though they all may present similar symptoms, the allergens may be different in each one. When the allergens were determined, and eliminated through NAET, they were free of their symptoms.

Freedom from Environmental Sensitivities

I would like to write one symptom for each different commonly seen environmental allergen here but this book will be too long. but that would give the reader some idea how contactants affect people to produce symptoms often diagnosed as "incurable," or " learn to live with it." We know there are many people suffering from these problems who spend their entire life visiting doctors and taking medications, still not finding any relief. I was one of them before I found NAET. If you can learn to identify the cause, at least one can avoid them and that itself will produce relief. For their sake, I am going to list some of the problems below and next to it in brackets the name of an allergen that has caused such a problem in a patient who got relief after the allergy to the particular allergen was eliminated. This may assist readers to search for culprits that just might be seemingly innocent contactant they are touching every day.

Angina-like pains (nightgown)

Anxiety (Dry cleaned blouse)

Arthritis on the wrist and finger joints (Tennis balls)

Asthma (Cotton shirt, cold weather, snow, dampness)

Bronchitis (Sweat suit)

Carpal tunnel syndrome (Computer mouse)

Chest pains (Body lotion)

Chills (Fleece material)

Constipation (Cotton underwear)

Coughing in bed (Down Pillow)

Cramps in the legs (Socks)

Cramps in the lower abdomen (Sweat suit)

Crying spells (Cooking smell)

Depression (Smell of seasoning and of coffee)

Dizziness (Newspaper)

Eczema (Fabric, body lotion, talcum powder)

Fainting spells (Reading newspaper)

Fatigue (Fabrics, Jewelry, computer)

Fibrocystic breast (Bra)

Fibromyalgia (Fabrics, Jewelry)

Frequent urination (Cotton underwear, toilet tissue)

Hay fever (Wood smoke, facial tissue)

Head heavy and pain all over the head (Tap water, drinking water)

Hives (Upholstery, fabric, dust, animals)

Insomnia (Bed linen, television remote control, bedside table, alarm clock)

Itching (Fabrics, jewelry)

Joint pains (Fabrics, jewelry)

Lower back ache (Elastic)

Mental confusion (Humidity)

Mental fog (Newspaper)

Mental irritability (Heat, jewelry)

Mid backache (Elastic, metal)

Migraine headaches (Fabrics, flowers, grasses, smells)

Freedom from Environmental Sensitivities

Muscle spasms (Fabrics)

Numbness on the foot (Shoe)

Pain anywhere in the body (Fabrics)

Pain in the cheek (Cold)

Pain in the hip (Elastic, fabrics, shoes, nylons)

Pain in the knee (Fabric, socks, shoe, walking cane)

Pain in the ankles (Socks, bathroom slippers)

Pain in the calf muscles (Knee highs, socks, pants)

Paranoia (Fabrics)

Pin prick pains (Fabrics)

Prostatitis (Cotton fabrics)

Pain in the sole of the foot (Bathroom slippers)

Sciatic neuralgia (Name tag on the pant, pants, elastics)

Shivering (Fabrics)

Sinusitis (Grass, pollen)

Skin cancer (Sun radiation)

Skin rashes (Sun, heat, fabrics, water)

Sneezing attacks (Smell from flowers, seasoning, cook ing)

Stiff neck (Jewelry, high neck-fabrics, Paper, pen)

Stomach aches (Cold, Fabrics, cooking, touching food products during preparation)

Swelling of the body (Water, cold)

Tennis elbow (Tennis racket, ball, glove)

Uterine fibroids (Humidity, dampness)

Various kinds of arthritis (Weather change, fabrics)

Vulvodynea (Underwear, elastics, fabrics, oxalic acids, berries, wood furniture, body lotion with vitamin C products)

A woman, 23, suffered from a type of dermatitis that did not respond to any traditional treatments. In our office she was treated for food products, chemicals and environmental items. Various parts of her body, such as neck, lips, face, chest, etc., showed dry, cracked furrows with clear water-filled blisters which looked like first degree burns. Often she had blood and serum oozing from weeping blisters. History revealed that she had been fine until she moved in with her fiance, a year and half before. She was allergic to her fiance's saliva. Her mysterious problem was solved when she was treated for his saliva by NAET.

John, 38, had suffered from depression for most of his life. He had tried various treatments, including psychotherapy, exercise, mega-nutrition therapy, massages, etc. He was found to be allergic to iron. He had wrought-iron ornamental works all over his house. When he was treated for the metal iron, his depression cleared.

Another woman, 28, was under treatment for lupus at a lupus clinic. She had severe joint pains most of the time. She suffered from severe insomnia, mental cloudiness, poor memory, mental irritability and debilitating multiple joint pains. She was on three different kinds of analgesics, which she took every three hours, to control her pain. Extremely hot, cold or cloudy weather affected her immensely. On such days, she stayed indoors with pain pills and warm water. When she was evaluated in our office, she was found to be allergic to all the fabrics she was wearing; however, she was not allergic to any food or drugs. She was treated individually for cotton, polyester, acrylic, nylon, plastics and leather.

At the end of the sessions, her lupus symptoms had diminished greatly. Her bodily disturbances with the weather changes also disappeared. She had been visiting the lupus clinic once a month. When she visited the clinic after she cleared her allergies, her blood test showed great improvement. She was told the best news by her doctor–that her lupus was in remission. Six years later, she remains absolutely symptom-free.

Woolen clothes may also cause allergies. We have seen people who cannot wear wool without breaking out in rashes. Some people who are sensitive to wool also react to creams with lanolin base, since lanolin is derived from sheep wool. Some people can be allergic to cotton socks, orlon socks, or woolen socks with symptoms of knee pain, etc. People can also be allergic to carpets and drapes that could cause knee pains and joint pains.

We had a few other female patients who were allergic to their panty hose and suffered from leg cramps, high blood pressure, swollen legs, psoriasis, and persistent yeast infections. Toilet paper and paper towels also cause problems, mimicking yeast infections in many people.

Other career-produced allergies have been diagnosed for cooks, waiters, grocery-store keepers, clerks, gardeners, etc. Virtually no trade or skill is exempt from contracting allergens.

Bill, 49, complained of severe pains in the right elbow, the wrist joint and the first interphalangeal joints. He had been treated for carpal tunnel syndrome, tenosynovitis, and tennis elbow many times before he came to us. When he was evaluated in our office, he was found to be highly allergic to paper, one of the tools of his trade as a writer. His symptoms cleared up soon after he was treated for paper successfully with NAET.

Another example of a paper allergy was observed during an interview with an attorney, who complained that he always came away from his office with a headache and feeling so tired that he could only go home and immediately to bed. This attorney was allergic to paper, with a completely different reaction from that of the writer.

Mary, a piano teacher with acute asthma, complained of wheezing every time she played the piano. It was discovered she was allergic to the ivory of the keyboard.

Ray, a salesman who complained of severe backaches after being on the road for a day, was found to be allergic to the acrylic seat covers in the car.

ALLERGY TO JEWELRY

People can be allergic to the jewelry and clothing they wear and sometimes to the jewelry and clothing others are wearing. They may also be allergic to such items when visiting friends, shopping, attending theaters or going anywhere people congregate.

For ten years, a young man had constant yeast infections, chronic fatigue syndrome, emotional fatigue, night sweats, nervousness, poor memory and various mental disturbances. He had seen a number of medical doctors, chiropractors, acupuncturists, and nutritionists to get some relief. Often suicidal, he was going for psychiatric counseling regularly. He was treated in our office for various allergies for over six months and showed marked progress.

He stopped having night sweats and fatigue, his memory improved and the yeast infection cleared up for the first time in ten years. He began to live normally once again. He was very happy and had a new girlfriend for a few months, when all of a sudden he returned to the office in tears. He said he was almost back to where he had started. We tested him for the various items that he was once treated for and found no allergy to anything. We noticed that he was wearing a lot of jewelry: four earrings, with four different stones in one ear, a heavy necklace with a huge gold and silver pendant, eight rings on the fingers with stones (star ruby, diamond, emerald, garnet, turquoise, sapphire, etc., and a gold watch studded with diamonds). When he was tested for the stones, he was found to be highly allergic to them. He revealed that he had taken off all the jewelry when he was treated for minerals in our office and did not put them back until recently. He was fond of jewelry and for ten years had always adorned himself with these jewels. Since he was feeling better, he started wearing them again and all the previous symptoms returned. He was treated for all the jewelry and once again he became healthy and happy. Moderation is the one word that can keep everyone out of agony.

CRYSTALS AND STONES

Many times, stiffness, pain in the shoulders and upper back tension can be the result of an allergy to jewelry.

A young woman who was a patient in our office complained of getting sinus blockages, pain in the upper arms and sometimes numbness on certain parts of the body whenever she sat in our waiting room. She claimed she felt better away from the office. The office staff felt guilty about this comment and decided to investigate.

She was scheduled for treatment at the same time as another woman who always wore a large crystal pendant. They were both in the office at the same time waiting to see the doctor. We discovered that the first woman was allergic to the crystal the other woman was wearing. The energy field of the crystal was very strong, like diamonds, and was affecting the first patient's field from ten feet away. From that day on, these two patients were scheduled at different times until the patient with the allergy to the crystal could be treated successfully.

Gold has been used in treatment for arthritis for years. If you are allergic to gold you can experience the exaggerated symptom of arthritis. Whenever you are allergic to a substance, it can generate the exaggerated symptom of the disease that was supposed to have been helped by the same substance.

A 27-year-old female was a victim of an allergy to gold. She was married at age 22 and, according to her custom, a tradition in India, she wore a thick gold chain around her neck. She was not allowed to remove the chain until she or her husband died. She began to have severe neck and shoulder pains. She went from doctor to doctor for some relief. One day her pain got so severe that she had to be rushed to the emergency room. X-rays of the neck showed a clean cut of the C5 vertebra exactly beneath the chain. She was kept in the hospital, and the vertebra was fused. She was all right for two months. Then she began to have the excruciating pain, all over again. This time she came to us, and we found that she was allergic to the thick gold chain she was wearing. She was treated for gold, and her neck pains and body aches disappeared. She still wears the chain around her neck. Since her treatment for gold, she does not get any more pains or other unpleasant reactions when she wears her gold necklace.

SICK BUILDING SYNDROME

Formaldehyde is found in fabrics, newspaper, ink, pressed wood, sheet rock, building materials, new carpet, name tags on clothes, paints, etc. Many people have been diagnosed as suffering from "sick-building syndrome." People who work in the newspaper industry, and writers suffer from carpal tunnel syndrome, which is often an allergy to formaldehyde or plastic.

Other career-produced allergies have been diagnosed for cooks (spices, cooking oil smells), waiters (smell of the foods, plastic trays), grocery store employees or workers (various items, chemicals to fresh food found in the store), clerks (paper products, inks, liquid paper, permanent markers), gardeners (pesticides, herbicides, plants, tools, leaf-blowers), computer programmers (computer radiation, plastics, electrical cords, mouse, keyboard, monitor), teachers (white board, pens, markers, paper, glue), bakers (flour, eggs, baking powder, artificial colors, flavorings, food chemicals), surgeons (latex gloves, gowns, masks, surgical instruments, antiseptics), lawyers (paper, pens, markers, computers, books, leather chair, wood-office furniture, cell phones), etc. Virtually no trade or skill is exempt from being exposed to contactant allergens.

INJECTANTS

Allergens are injected into the skin, muscles, joints and blood vessels in the form of various serums, antitoxins, vaccines and drugs. As in any other allergic reaction, the injection of a sensitive drug into the system runs the risk of producing dangerous allergic reactions. To the sensitive person, the drug actively becomes a poison, with the same effect as an injection of arsenic. No one would intentionally give an injection of a potentially dangerous drug to a person. However, some drugs seem to become more allergenic for certain people over time, without the person being aware of the potential risk. Take the increasing incidents of allergies to the drug penicillin as an example. The reactions vary in people, from hives to diarrhea to anaphylactic shock and death.

Most of us do not often consider an insect bite in the same way as we would an injection received from a physician or a member of his staff, but the result can be quite the same. At the point of the bite, a minute amount of the body fluid (saliva) of the insect is injected into the body. This fluid is formulated to produce immobilizing pain in order to protect the insect from its own predators.

Many people react to bites from insects. Spider and flea bites can be fatal in some cases. Many people react to wasp and bee stings. This case study is taken from another NAET doctor's practice: One of his patients came in with many spider bites on her arms. She was bitten while working in her vegetable garden. The bites were very painful and her arms were swollen. The doctor asked her to bring one dead spider from her garden. She was treated by NAET for allergy to the spider. After treatment, her wounds healed, and she was no longer bothered by spider bites from her garden.

Flea bites are another commonly ignored allergic item for many allergy sufferers. A nine-year-old girl, who had severe asthma for eight years, was treated for various allergies, bringing her asthma under control. After she was treated for animal dander, cats and dogs, her asthma did not bother her any more. One day she visited some of her friends who had a dog. After a few minutes, her asthma returned. She was allergic to the fleas. When she was treated for the fleas, she stopped having breathing difficulties when she goes near cats and dogs.

A woman, 59, came to the office with a typical case of accidental serum injection poisoning. She was vacationing in New York where she was stung by a bee while boarding a bus. Although she had known that she reacted strongly to insect bites as a child, she had no idea how much the allergic condition had increased over the years. Within minutes, she was feeling nauseous and light-headed and was having difficulty breathing. Luckily, she was taken to a hospital emergency care unit. By the time she reached the hospital, she experienced some respiratory distress. She was treated by the doctors and hospitalized for three days.

When she returned to California ten days later she was still experiencing some cellulitis in her left arm. She was brought to our office. After she received treatments by NAET to desensitize her for bee sting, the cellulitis in the arm diminished. One year later, when she was camping, bees stung her again. She panicked and her friends drove her to the nearest hospital, about 40 miles away. They sat in the hospital parking lot and waited for two long hours to see whether she was going to have any reaction. Since she felt all right, they returned without entering the hospital. The next day she came to our office. She had many bee sting marks on her arms, which looked like mild prickly heat, but never experienced any unpleasant reactions.

Most of our NAET patients are well trained to take care of their own unexpected emergencies if they encounter any situations when they are away. By the time they complete their Basic 15 groups, they will be taught self-balancing techniques through our patient-educational seminars.

Caroline was a chronic fatigue syndrome patient. She was seeing us for almost a year and was doing well overall. Her husband also suffered from numerous ailments and continued taking various medications without much result. She suggested that he try NAET but he didn't believe in NAET. He didn't think tapping on certain parts of the body could eliminate an allergic reaction that a powerful pharmaceutical druig could not eliminate so far. He told his wife that he was not going to be fooled by anyone like that. As for her, he said she was too gullible. Since her mother was paying for her treatment, he had nothing to say. She was very emotional about this. Not only did it hurt her watching her husband suffer from his ailments, he also ridiculed NAET, the only treatment that helped her to overcome the fatigue she suffered for years. I gave her an emotional balancing treatment to reduce her emotional pain.

A few weeks later, she and her husband decided to do some work in their vacant rental unit. They were getting the place ready to rent. He was clearing the weeds and trimming the trees when a bee stung him. He had experienced severe life-threatening reaction to bee stings on two previous occasions. Both times he had to be rushed to the hospital. Caroline was standing near him when this happened. He panicked and told her to get emergency help. She didn't have a phone with her. No neighbor was home.

Immediately she thought about NAET. She looked at her husband. He was flushed, his face looked red beet. He was breathing fast– hyperventilating. She ran to him, made him lie on the concrete driveway, and began massaging his GV 26, (the acupuncture cardiopulmonary resuscitation point - Read Chapter 7). In a few seconds his breathing became easier. Suddenly she saw a bee lying on the grass unable to fly. She quickly grabbed it with a tissue, wrapped it inside the tissue, made him hold the tissue, and massaged his NAET home-help points. She massaged them one after another clockwise, one minute on each point. When she completed one round of massaging all eight points, she continued to massage again all over. While she was massaging the points, she watched his face. When she completed three rounds of massaging the points, she saw his face relaxing and the color returned to normal. She massaged the points nonstop for 45 minutes and while massaging he fell asleep. When she stopped the massage he woke up with a smile. He was breathing normally. He had a minor mark which looked like prickly heat on his forearm, but no other symptoms. He sat up gave his wife a hug and said, he was now a believer in NAET.

The normal reaction to a bite, other than lethal bites, ranges from mild swelling around the site of the injection, a mild reddening and, often a slight to moderate discomfort in the body from attempting to free the toxin that produces itching. Rarely are these bites and stings lethal to the normally non-reactive person.

For some people, however, a sting or a bite by an animal or insect can be potentially lethal. Even a single mosquito bite may produce an extreme and sudden onset of edema (the abnormal collection of fluids in the body tissue and cells) and severe respiratory distress. There have been many cases of anaphylactic shock, respiratory and/or cardiac failure in sensitive persons following the slightest insect bite.

Various vaccinations and immunizations may also produce such allergic reactions. Some children after they receive their usual immunization get very sick physically and emotionally. We have seen many children suffering from autism and attention deficit and hyperactive disorders after receiving immunizations. It is not the immunizations, per se, that cause the problems, but allergy to the ingredients in the vaccination that cause havoc in some children and adults.

INFECTANTS

Infectants are allergens that produce their effect by causing a sensitivity to an infectious agent, such as bacteria, virus, or parasite. For example, when tuberculin bacteria is introduced as part of a diagnostic test to determine a patient's sensitivity and/or reaction to that particular agent, an allergic reaction may result. This may occur during skin patch, or scratch tests done in the normal course of allergy testing in traditional Western medical offices.

It should be noted that bacteria, virus, etc. are contacted in numerous ways. Our casual contact with objects and people exposes us daily to dangerous contaminants and possible illnesses. When our autoimmune systems are functioning properly, we pass off the illness without notice. When our systems are not working at maximum performance levels, we can experience infections, fevers, etc.

Maxine, 72, had chronic weeping ulcers on the tips of the fingers of both her hands for many years and had tried all the possible medicines and ointments. While taking her history, it was revealed that caring for roses was her main hobby. Having taken this into consideration, she was tested for roses by NST and was found to be highly allergic to them. When she was treated by NAET, her chronic ulcers healed nicely.

PHYSICAL AGENTS

It has been recognized for many years by various physicians that extremes of temperature, particularly of cold, may bring on attacks of asthma or hives. As long ago as 1860, a physician described asthma attacks produced in a man after applying cold water to his instep. In 1866, another physician reported that cold water could produce intense hives on the skin. In 1872, another physician reported the case of a 45-year-old woman whose hand would swell when immersed in cold water. The swelling subsided 15 minutes after reaching its peak. Cold air produced hives on her face and neck. Eating ice cream immediately caused her intense pain in her throat and a feeling of suffocation.

Weather changes can greatly affect many allergic patients. People with arthritic and respiratory tract problems are affected by high or low temperatures. When outside temperature falls, the humidity increases and affects asthmatics, in some cases even becoming fatal. Change in altitude can also cause difficulties for allergic people. Some people do well in high altitudes, whereas others cannot tolerate them at all. In patients with problems in high altitudes, treating with NAET for exhaled air can give relief.

Some people are so severely affected by the cold that they faint or lose consciousness. Numerous deaths from drowning may have been caused by an allergic reaction to cold. The swimmer becomes immobile and/or unconscious due to the severity of pain.

As in other types of allergic responses, physical allergy may be manifested as:

- Hives or angioneurotic edema of the hands when washed in cold water, and hives on areas of the body exposed to cold air.

- Swelling of the lips or spasms of the stomach after eating cold foods.

- Allergic rhinitis and asthma may be brought on by the inhalation of cold air (responds well to the NAET wind and dampness treatment).

- Skin turning blue in cold air and red under a warm sun.

Low concentrations of certain nutrients can throw the body into imbalance and extreme discomfort when the weather changes. Lack of iron, vitamin B 12, and folic acid may cause poor blood quality and inadequate circulation, blocking the body's ability to exchange oxygen and carbon dioxide properly then adding more difficulties in extreme temperatures. Many patients who have extremely cold feet or cold hands suffer from poor circulation. After treating for iron with NAET and then taking iron supplements, their circulation improves.

There are extreme cases of patients with physical allergies. One patient complained of pains in the mouth, esophagus (the canal connecting the throat and stomach), and the stomach after drinking cold water. Cold air caused her lips and tongue to swell, made her eyes water, and caused coughing. Prolonged exposure to cold caused asthma. She was treated for basic allergies to all the essential nutrients such as calcium, iron, vitamins C and A, trace minerals, salt, B complex vitamins, sugars, and fats. By the end of these treatments she no longer suffered from an allergy to cold. When she cleared her allergy to ice cubes, she was even able to drink cold water and eat ice cream.

Another patient, a girl who was a victim of diabetes and kidney failure, could only suck on ice cubes to quench her thirst. (She was not allowed to drink water.) She developed chronic bronchitis and a cough that did not respond to a series of strong antibiotics. After almost eight months of chronic cough, we tried to treat her for ice cubes. It took four consecutive NAET treatments to clear the allergy. To our amazement, she cleared the cough and chronic bronchitis upon passing the ice cube treatments.

Another patient suffered from hives, itching, and redness of the skin, headache, diarrhea, general weakness, and fainting spells whenever she walked in the hot sun, drank hot liquids or took a hot or warm shower. She was treated for hot water and has not reacted nearly as much to heat since. A woman suffered from severe asthma only in the summer, but not in winter. When she was treated for food allergies and combination with heat her asthmatic reaction to heat was reduced.

Many people working with computers are having various health problems arising from an allergy to computer radiation or an allergy to keyboard, mouse, or the computer itself. The symptoms arising from computer work include: eczema, carpal tunnel syndrome, pain in the elbow, shoulder, sometimes in both sides of the body eventhough they may be using computer or mouse by one hand only, fatigue, fibromyalgia, excessive sweating, poor appetite, chest pain, chronic nonproductive cough, insomnia, weight gain, constipation, water retention, itching, hives, nervousness, panic attacks, seizures, etc. In such cases, after all the NAET Basics, one needs to treat soft and hard plastic, radiation, heat, one's own computer keyboard, mouse, screen, etc.

Very often, physical allergies may exist in patients who are also allergic to other substances. While the physical allergy in itself is not sufficient to bring on an allergic attack, contact with the offending allergen plus the physical agent may precipitate an attack. An example of this type of allergy was seen in a woman who frequently suffered asthmatic attacks produced by sudden exposure to cold wind. When she was treated for the foods she was sensitive to, exposure to cold wind also lost its effect. Sometimes people are allergic to both heat and cold, and this makes them miserable all the time. Such was the case of a man who developed hives and edema, and sometimes dizzy spells, after he played tennis. When he took cold showers, he had similar complaints. He was also found to be allergic to: tennis balls, tennis rackets, shoes, soaps, shampoos, various foods, etc. After he was treated for his allergens, as well as for hot water and ice cubes, he no longer reacted to heat or cold.

Some people are sensitive to the sun's heat. In this case, they may be allergic to the sun's actinic rays only. These people are also allergic to various other items. In the presence of other allergies, an allergy to the sun's rays brings on worse reactions. Such was the case of a man who worked on a farm and had severe dermatitis, with redness, extreme swelling and itching of the face, lips, neck and hands. He could only work during the early part of the morning hours or late in the evenings. The rest of the time he spent indoors. He was also found to have various food allergies. After treatments for all the food items, his reactions to the sun were also reduced.

When physical allergy occurs in conjunction with pollen, foods or other substances, it is possible to treat the patients successfully by treating for the other nonphysical allergens. Most of the time, patients do not continue to react to the physical agents. When people suffer from allergy to physical agents only, they must be treated for cold or heat. Such cases are very rarely seen. In these rare cases, when no other treatments work, it is wise to avoid heat-producing foods (according to Chinese nutrition theory - sugar, fat, etc.) in heat-sensitive patients and cold-producing foods (rice, starches, ice cream, etc.) in cold-sensitive patients. The author has not seen any cases in her practice that did not respond well to some kind of allergy treatments.

Joseph, 67, came to our office with complaints of excruciating pain and red rashes on the lateral parts of the thigh, leg and foot for more than a week. He had seen a few medical professionals for this problem without any result. His pain and immobility kept increasing. NAET testing determined an allergy to the sardines he had eaten a week before. He was treated for sardines by NAET, and his pain and rashes disappeared in a few hours.

Margarite, 59, came with a history of vitiligo (white spots all over the body, some under the chin, in front of the ears, leg, back of the neck) for almost 30 years. She was found to be allergic to wool, vitamin C, dust mites, yeast, molds, fruits, yogurt, vinegar, chlorophyll, whole grains and melanin. When she was treated for these items, her white spots were replaced by normal skin pigments.

Heat, cold, sunlight, dampness, humidity, dryness, moist heat, drafts or mechanical irritants may also cause allergic reactions and are known as physical allergens. When the patient suffers from more than one allergy, physical agents can affect the patient greatly. If the patient has already eaten some allergic food item, then walks in cold air or drafts, he might develop upper respiratory problems, sore throat, asthma or joint pains, etc., depending on his tendency toward health problems. Some people are very sensitive to cold or heat, whether they have eaten any allergic food or not. Such cases are common.

One of the young patients who came to our office had a history of canker sores whenever he walked in the sun. He was highly allergic to vitamin D, one of the vitamins produced in the body with the help of sunlight. After he was treated by NAET for vitamin D, the incidents of canker sores as a result of walking in the sun diminished.

Esther, 42, became very disturbed mentally and physically whenever it was hot. Her whole body swelled up in the hot season or if she walked in the sun for a few minutes. She could never drink hot liquids or eat hot foods without getting sick. Whenever she ate hot foods, she would have a paroxysmal tachycardia (very speedy heart rate) and her whole body would turn red and swollen. She was treated for very hot water. A few minutes after treatment by NAET, she had severe abdominal cramps and severe continuous diarrhea. We had to treat her several times and fianally her body began to adjust to the heat in a normal way. A week after the treatment, there was a heat wave in Southern California. The temperatures reached 104 degrees. Her car broke down on a mid-afternoon, and she had to walk for about 15 minutes to reach a gas station for help. Amazingly, she did not react to the heat like she would have before the treatment. She did not even have the usual swelling of feet. Now she enjoys hot herbal teas and soups, and she has lost 35 pounds. In her case, the allergy to heat was the cause of her obesity.

Maria, 49, had experienced severe hot flashes for three years. She was on hormone supplements, but nothing gave her relief. She was found to be allergic to heat, sugar and hormones. After she was cleared for the allergy to the above items, her hot flashes stopped completely.

Jenny, 58, suffered from Raynaud's disease. The tip of her fingers remained dark blue on a cold day. She was allergic to cold, citrus fruits, and meat products. She felt better when she was cleared for the above items.

Many arthritic patients, asthma patients, migraine patients, PMS patients, and mental patients have exaggerated symptoms on cold, cloudy or rainy days. These types of patients could suffer from severe allergy to electrolytes, cold, or a combination of both.

Some patients react to heat or cold violently, getting aches and pains during a cloudy day, and icy cold hands and feet even if they are clad in multiple warm socks. These patients have hypo-functioning immune systems. When they finish the NAET treatment program, they do not continue to feel cold or get sick with the heat or cold.

John suffered from a continuous cough all winter (from October through April). When the weather warmed up, his cough got better. Every year, he went through the same problem. He drank cough syrup by bottles without any result. His friend referred him to us. When he was evaluated through NAET, he was found to be allergic to cold. He was treated for cold (ice cubes). For the past five winters in a row he has not had a cough.

Forty-nine-year-old Frank suffered from asthma all his life. As he was getting older, his attacks intensified. Inhalers and cortisone refused to work on him as before. He was forced to increase the dosage frequently to get some results. Then he heard about NAET through one of his friends. Another NAET specialist treated him for various food and environmental substances. His energy improved and he felt better overall, but his asthma did not get better. After 84 treatments, his practitioner referred him to me for evaluation. By evaluating through NAET testing procedures, I found that he was reacting to dampness. He was treated for dampness immediately. In less than ten minutes, he reported that his lungs and respiratory passages opened up. "Something lifted off my chest," he exclaimed. "I can feel the air filling up my lungs for the first time in my life!" After that treatment, he did not have asthmatic attacks.

SEASONAL AFFECTIVE DISORDERS

Many people suffer from depression during the winter season. They seem to be reactive to the long nights (darkness). They feel better when the summer returns. This is a problem for people living in places like Alaska where they have long, dark nights during winter. Some people feel better if they sit in front of bright white lights for a few hours everyday. Many people begin to have food cravings during winter. Such people may need to get treated for vitamin D, calcium, pineal gland, sugar, serotonin, fatty acids, magnet, and salt. NAET treatments can help with this condition.

Seana, 42, suffered from winter depression and became suicidal. She was found to be allergic to vitamin D, calcium, sugar, serotonin, melatonin, magnet, pineal gland, and fatty acids. After she was successfully treated for her allergies, she no longer suffered from her usual winter depression.

A 38-year-old woman had second-degree burns with huge blisters as a result of a container of boiling water falling on her. After one week of extreme pain, she was treated by NAET with very hot water in a glass jar. After 24 hours the pain was gone. Within a week, her skin was healed.

Sarah, a 64-year-old patient, visited Colorado Springs. At that high altitude, she noticed a shortness of breath and began sighing deeply. A local NAET practitioner treated her for carbon dioxide blown into a paper bag. Within a few hours, her breathing difficulties ceased. She was able to stay there for another week with no further problem.

LATEX ALLERGIES

To most people, latex gloves can be very helpful items but many people are allergic to latex gloves, one of the most common items found in hospitals. Hardly any procedure is done in the hospitals without latex gloves. If the allergic person happens to use them, it can be life threatening.

A 48-year-old woman went to the dentist for a routine cleaning. The dental hygienist put on a pair of latex gloves and began cleaning her teeth. All of a sudden, her body became burning hot. The patient almost went into anaphylactic shock. The doctor came in and called the paramedics who revived her with the help of drugs. She had red rashes all over her body and a mild fever when she left the dentist's office. She came to our office two days later with red rashes still on her body and a fever of 100 degrees Fahrenheit.

Testing through NTP showed she was allergic to the latex gloves the dental hygienist used. She was also allergic to the chalk powder in the gloves, which affected her large intestine, heart and gallbladder meridians. She was treated for the gloves and the powder. After she was treated for the gloves, she was also treated for local anesthetics, amalgam, cleaning agents, gauze and the cotton balls. She was advised to return to the dental office to finish up the dental work. This time, she had no adverse reaction. She even had a root canal done. Later, she revealed that the allergy elimination treatment for latex gloves and powder was the best thing that had happened to her in 15 years. She had been afraid to have any intimate relationship with her husband for many years. She reacted to his semen and broke out in rashes and blisters all over her body whenever they had intercourse. Actually, she was allergic to the condom materials and couldn't use them or be around them. Later, she was successfully treated for her husband's semen, and it goes without saying that she is a very happy woman now.

Mike, 32, who frequently had skin cancer on his face, was evaluated and found to be highly allergic to the stainless steel blade of the razor he used. The use of a popular skin cream was the cause of the beginning of skin cancer in another patient. She was allergic to vitamin A and to the skin cream itself. After she was treated for these substances, her lesions cleared up.

GENETIC CAUSES

Discovery of possible tendencies toward allergies carried over from parents and grandparents opens a large door to achieving optimum health. Most people inherit the allergic tendency from their parents or grandparents. Allergies can also skip generations and be manifested very differently in children.

Many people with various allergic manifestations respond well to the treatment of various disease agents that have been transmitted from parents.

Jill, 55, suffered from Epstein-Barr virus and various allergies. After treatment for the virus, her response was very encouraging. Upon questioning her, it was found that her Japanese parents, uncles and aunts died of tuberculosis. She was immediately tested and treated for tuberculosis and she became allergy free and healthy once again.

Parents with rheumatic fever may transmit the disease to their offspring, but in the children the rheumatic fever agent may not be manifested in its original form. As an example, Sara, 42, had severe migraines all her life. Her mother had rheumatic fever as a child. Treatment for rheumatic fever lessened her migraines.

MOLDS AND FUNGI

Molds and fungi are in a category by themselves, because of the numerous ways they are contacted as allergens in everyday life. They can be ingested, inhaled, touched or even, as in the case of Penicillin, injected. They come in the form of airborne spores, making up a large part of the dust we breathe or pick up in our vacuum cleaners; fluids such as our drinking water; as dark fungal growth in the corners of damp rooms; as athlete's foot, and in particularly obnoxious vaginal conditions commonly called "yeast infections." They grow on trees and in the damp soil. They are a source of food, as in truffles and mushrooms; of diseases such as ring worm and the aforementioned yeast infections, and of healing, as in the tremendous benefits mankind has derived from the drug Penicillin.

Reactions to these substances are as varied as other kinds of allergies. This is because they are a part of one of the largest known classifications of biological entities. Because of the number of ways they can be introduced into the human anatomy, the number of reactions are multiplied considerably. Fungi are parasites that grow on living as well as decaying organic matter. That means that some forms are found growing in the human anatomy. The problem of athlete's foot is a prime example.

Athlete's foot is a human parasitic fungus that grows anywhere in the body, where the area is fairly moist and not exposed to sunlight or air. It is particularly difficult to eliminate, and treatment generally consists of a topical preparation, multiple daily cleansing of the area, a medicinal powder, and wearing light cotton socks to avoid further infection from dyes used in colored wearing apparel.

It is contracted by contact with the fungus and is often passed from person to person anywhere there is the potential for contact (i.e.gymnasiums, showers, locker rooms and other areas where people share facilities and walk barefoot), thus the name athlete's foot. If it is a real athlete's foot, it will clear with the NAET treatment. Certain allergies, like allergies to socks made of cotton, orlon, or nylon, etc., can mimic athlete's foot. In such cases, athlete's foot may not clear by using medications.

Allergies to cotton, orlon, nylon, or paper could result in the explosions of infections including Ascomycetes fungi (yeast) that women are finding so troublesome. Feminine tampons, toilet papers, douches, and deodorants can also cause yeast infections.

Waterbeds are comfortable places for molds to grow when the user forgets to add chemicals regularly into the water inside to inhibit growth of molds. Users must also be careful to test the chemicals for possible allergic reactions. We have found people getting bronchitis that lasted a year in spite of taking repeated antibiotics, chronic sinusitis, migraines, back aches, etc. when sleeping in untreated waterbeds.

EMOTIONAL STRESSORS

Many times, the origin of physical symptoms can be traced back to some unresolved emotional trauma. Each cell in the body has the capability to respond physically, physiologically and psychologically to our daily activities. When the vital energy flows evenly and uninterrupted through the energy pathways (acupuncture meridians), the body functions normally. When there is a disruption in the energy flow through the meridians (an increase or decrease), energy blockages can occur, causing various emotional symptoms in those particular meridians. According to Oriental medical theory, there are seven major emotions which can cause pathological health problems in people: sadness affects lung meridian, joy affects the heart, disgust affects the stomach, anger affects the liver, worry affects the spleen, fear affects the kidney, and depression affects the pericardium meridian. Please read Chapter 10 in Say Good-bye to Illness for more information on meridians and emotional connections.

Cynthia suffered from asthma since she was nine years old. She was 15 when her mother brought her for NAET treatments. According to her, she had asthma only when she played or sat on the grass. If she never came in contact with any grass, she would go without getting asthma for months. On evaluation we found her allergic to all Basic fifteen groups. After completing the basics we decided to treat her for NAET sample grass mix. As soon as I treated her for grass mix, she complained of severe abdominal pain, sat on the floor, doubled over in pain. I tested her through a surrogate and she was passing grass at the physical level but the emotional level had failed. NTP revealed she had an issue with fear that happened when she was nine years old. I isolated the issue with a male, non-family member. Cynthia or her mother couldn't remember anything that happened to cause such fear in her at that age. I treated her for fear randomly and her pain reduced. She was still getting queasy in her stomach. They thought eating some food might settle her stomach and went to a nearby restaurant. As they walked in, Cynthia saw a man with a long beard passing by. She let out a cry and she felt her abdominal pain returning. Puzzled, her mother inquired.

Cynthia said, "That man reminded me of something and I am feeling sick again."

Then her mother remembered the story. When Cynthia was nine years old she and two of her other friends playing on the front lawn of the house on a summer day. It was very hot that day. Cynthia was lying on the cool grass. Others were sitting. A man with a long beard, dirty clothes and a large empty black bag walked towards the children mumbling something. The previous week their school had warned the children to be aware of a man with beard, dirty clothes and a black bag. There was a rumor that he was a child-snatcher. Both of her friends yelled, "Here is the child-snatcher, run, run!" and they ran into the house. Cynthia tried to get up from her lying position and fell on her knees in panic but somehow managed to get up and run before he caught her. Hearing the noises Cynthia's mother came outside and found the man ringing the door bell. He was the newspaper delivery man coming to collect the dues. She now remembered how frightened Cynthia was about men with beards after that incident.

They immediately returned to the office and shared the story. I treated her for that fear. Ever since, Cynthia did not have asthma on or near any grass. The NAET I gave her was able to turn off the fear-switch towards the fear she had towards all men with beards. along with that her brain turned off the switch towards getting asthma. The human brain is a fascinating instrument with hundreds of hidden secret keys to thousands of adventure lands.

ENVIRONMENTAL ILLNESS

Caused From An Emotional Blockage
From Childhood

I am a patient of Dr. Devi and Mala. I suffered from mild to moderate of depression and sadness for a long time. I always had this deep sadness and inner crying. Mala's testing indicated that the source was when some event happened somewhere around age 16, with both the emotional and chemical levels affected.

When I was 16 years old, I felt inferior economically because our family was very poor. I and a girl from a very wealthy family liked each other. We both belonged to the same church and youth group and attended swimming pool parties and all the other youth functions of the church. At the pool parties at the girl's house we did acrobatics. I picked her up by the hips next to the pool and balanced her horizontally over my head. We played volleyball in the pool with her sitting on my shoulders and did many other activities in the pool and had a lot of innocent fun.

When I reached home that night, I felt very sad for being poor and not equal to her. During the pool activities, I wanted to ask her for a date and I felt that she wanted me to ask but I couldn't because I was divided, in a dilemma, in a paradox, which came out of my feeling of economic inferiority, which, in turn, inhibited and intimidated me, so I never asked.

Mala brilliantly correlated that at the same time I was feeling conflicting, I was smelling chlorine and all the other pool chemicals. My mind and nervous system correlated and registered the sensation of emotional conflict/catatonia with the smell of chemicals.

For me, Dr. Devi Nambudripad definitely opened a whole new world of the real allergy elimination, thereby giving much hope for all generations to come. I thereafter developed an allergy to chemicals which progressively worsened. Later in life, whenever I smelled chemicals, I experienced various degrees of brain and body pain and catatonia. At certain times when I had the opportunity to meet a girl where there was a mutual attraction, I couldn't approach her.

A few days after Mala's combination emotional/chemical NAET treatment, the long term deep sadness has lifted 100%.

Dikran A, Ph.D.
Los Angeles, CA

3

Detecting Allergies

A detailed clinical history is the best diagnostic tool for any medical condition. It is extremely important for the patient, his/her parents or guardians to cooperate with the physician in giving all possible information to the doctor in order to obtain the best results. It is my hope that this chapter will help bring about a clearer understanding between NAET specialists and their patients, because in order to obtain the most satisfactory results, both parties must work together as a team.

Your doctor's office may ask you to complete a relevant questionnaire during your first appointment. It is important to provide as accurate a history as possible.

Patient's Data Sheet

This data is essential to understand the patient's background. It should contain the following information: age, gender, marital status, job situation, work address, home address, telephone numbers, other identifying information, and a next of kin to contact in emergency.

Review of personal data will help the allergist understand the patient better: Is the patient a child/ under-age? A student? or a mature adult? Employed or unemployed? Disabled due to illness? Capable of taking care of his/her needs, or dependent? Residential address, work address and job situation will help the allergist to understand living and working environments.

Social History

Learning: If the patient is a child: grades at school, interaction between friends and teachers, interaction between family members, activities at school, phobias, and problems with discipline.

If the patient is an adult: the appropriate social history.

Behaviors: If the patient is a child: cooperative, uncooperative, disruptive and/or aggressive behaviors; overactive, restless, inattentive, day dreams; uncooperative with peers or adults; incomplete or sloppy work.

If the patient is an adult: the appropriate history of behavior patterns.

Habits: If the patient is a child: temper tantrums, excessively active, constantly moving in seat or room, low self-esteem, short attention span, poor memory, unusual fears; falls down often, clumsy, and unintentionally drops things.

If the patient is an adult: the appropriate history of habits.

Hobbies: reading, painting, singing, walking, cooking, sports, riding, etc.

Personal Health History

The first step in diagnosing an allergy is for the allergist to take a thorough health history, including, among other things, a record of any allergic symptoms in the patient's family. The patient will be asked whether either of his/her parents suffer from asthma or hay fever, ever suffered from hives, reacted to a serum injection (such as tetanus antitoxin, DPT), or experienced any skin trouble. Additionally, the allergist will ask whether the patient or his/her parents were unable to eat certain foods; unable to go to places; did they have any environmental sensitivities? chemical sensitivities? complained of sinusitis, repeated headaches, runny nose, frequent colds or flu; had dyspepsia, indigestion, joint pains, mood swings, or any other conditions where an allergy might have been a contributing factor.

Family History

The medical history of the immediate relatives, mother, father, and siblings should be noted. The same questions are asked about the patient's relatives: grandparents, aunts, uncles, brothers, sisters, cousins. A tendency to get sick or have allergies is not always inherited directly from the parents. It may skip generations or manifest in nieces or nephews rather than in direct descendants.

The inquiry should include information on any family on alcoholism, autoimmune disorders, drug abuse, smoking, eating disorders, addictive behaviors, mental disorders, other health disorders. The careful NAET specialist will also determine whether or not diseases such as tuberculosis, cancer, heart disease, diabetes, rheumatic or glandular disorders exist, or have ever occurred in the patient's family history.

All of these facts help give the NAET specialist a complete picture of the hereditary characteristics of the patient.

Present History

When the family history is complete, the practitioner will need to look into the history of the patient's chief complaint and its progression. Some typical preliminary questions include: "When did your first symptom occur? When did you notice your child's first symptoms related to environmental sensitivities? Did you suffer from these problems since childhood? Did you notice your child's problem when he/she was an infant or child, or during adolescence? Did it occur after going through a certain procedure? For example, did it occur for the first time after a dental procedure, like applying braces or filling a cavity? Did it happen after the first antibiotic treatment or after installing a water filter? Did it occur after acquiring a water bed, a tricycle, or after a booster dose of immunization?" One of my patients reported that her son's asthma began a few days after he received a pet (puppy) for his birthday.

Next, the doctor will want to know the circumstances surrounding and immediately preceding the first symptoms. Typical questions will include: "Did you change your or the child's diet or put him/her on a special diet? Did you or he/she eat something that hadn't been eaten recently, perhaps for two or three months? Did you eat or feed your child one type of food repeatedly, say every day?. Did the symptoms follow a childhood illness, (whooping cough, measles, chicken pox, diphtheria) or any immunization for such an illness? Did they follow some other illness such as influenza, pneumonia, or a major operation? Did the problem begin after your vacation to an island, to another country, or after an insect bite? After eating a special food? After using a special chemical to clean your carpet? After taking a course of antibiotics? After starting a new vitamin supplement? After going through a detox program? After receiving other therapy? After an accident? After receiving traumatic news?"

Any one of these factors can be responsible for triggering a severe allergic manifestation or precipitate the first noticeable symptoms of an allergic condition. Therefore, it is very important to obtain full and accurate answers when taking the patient's medical history.

Other important questions relate to the frequency and occurrence of the attacks. Although foods may be a factor, if the symptoms occur only at specific times of the year, the trouble most likely is due to pollens. Often a patient is sensitive to certain foods but has a natural tolerance that prevents sickness until the pollen sensitivity adds sufficient allergens to throw the body into imbalance. If symptoms occur only on specific days of the week, they are probably due to something contacted or eaten on that particular day.

Many patients react violently to house dust, different types of furniture, polishes, house plants, tap water and purified water. Most of the city water suppliers change the water chemicals only once or twice a year. Although, this is done with good intentions, people with chemical allergies may get sicker if they ingest the same chemicals over and over for months or years. Contrary to traditional Western thinking, developing immunity can be the exception rather than the rule.

Occasionally switching the chemicals around gives allergic patients a change of allergens and a chance for them to recover from existing reactions.

The doctor should ask the patient to make a daily log of all the foods and activities he/she is eating or doing. The ingredients in the food should be checked for possible allergens.

When assessing a person, care must be taken not to misdiagnose. In my opinion, people who get labeled with various health disorders may be suffering from simple undiagnosed allergies. Many food and environmental allergic symptoms overlap or mimic a variety of diseases, including many neurological and brain disorders.

After completing the history, the NAET specialist should examine the patient for the usual vital signs. A physical examination is performed to check for any abnormal growth or condition. If the patient has an area of discomfort in the body, it should be inspected. It is important to note the type and area of discomfort and its relationship to an acupuncture point. Most pain and discomfort in the body usually occurs around some important acupuncture point.

NAET Evaluation Methods

NAET uses many standard allopathic and kinesiological testing procedures to detect allergies. Some of the common ones are mentioned below.

History

A complete history of the patient. A symptom survey form is given to the patient to record the level and type of discomfort he/she is suffering.

Physical Examination

Observation of the mental status, face, skin, eyes, color, posture, movements, gait, tongue, scars, wounds, marks, body secretions, etc.

Vital Signs

Evaluation of blood pressure, pulse, skin temperature and palpable pains in the course of meridians, etc.

SRT - Electro-Dermal Test-EDT

Skin Resistance Test for the presence or absence of a suspected allergen is done through a computerized electrodermal testing device; differences in the meter reading are observed (the greater the difference, the stronger the allergy).

NST

NST (Neuro Muscular Sensitivity Testing) is the body's communication pathway to the brain. Through NST, the patient can be tested for various allergens. NST is based on the principles of *Medical I Ching*, a 6,000 year old Chinese diagnostic modality. It is closely related to MRT (muscle response testing), a standard test used in Applied Kinesiology to compare the strength of a pre-determined test muscle in the presence and absence of a suspected allergen. If the particular muscle (test muscle) weakens in the presence of an item, it signifies that the item is an allergen. If the muscle remains strong, the substance is not an allergen. More explanation on NST and MRT will be given in Chapter 6.

Dynamometer Testing

A hand-held dynamometer is used to measure finger strength (0-100 scale) in the presence and absence of a suspected allergen. The dynamometer is held with thumb and index finger and squeezed to make the reading needle swing between 0-100 scale. An initial base-line reading is observed first, then the allergen is held and another reading taken. The finger strength is compared in the presence of the allergen. If the second reading is more than the initial reading, there is no allergy. If the second reading is less than the initial reading, then there is an allergy.

Pulse Test

Pulse testing is another simple way of determining food allergy. This test was developed by Arthur Coca, M.D. in the 1950's. Research has shown that if you are allergic to something and you eat it, your pulse rate speeds up.

Step 1: Establish your base-line pulse by counting radial pulse at the wrist for a full minute.

Step 2: Put a small portion of the suspected allergen in the mouth, preferably under the tongue. Taste the substance for two minutes. Do not swallow any portion of it. The taste will send the signal to the brain, which will send a signal through the sympathetic nervous system to the rest of the body.

Step 3: Retake the pulse with the allergen still in the mouth. An increase or decrease in pulse rate of 10% or more is considered an allergic reaction. The greater the degree of allergy, the greater the difference in the pulse rate.

This test is useful to test food allergies. If you are allergic to very many foods, and if you consume a few allergens at the same time, it will be hard to detect the exact allergen causing the reaction just by this test.

Hold, Sit and Test

This is a simple procedure to test allergies. Place a small portion of the suspected allergen in a baby food jar or thin-glass jar, preferably with a lid, then the person will hold it in her/his palm, touching the jar with the fingertips of the same hand for 15 to 30 minutes. If the person is allergic to the item in the jar, he/she will begin to feel uneasy when holding the allergen in the palm, giving rise to various unpleasant symptoms. This testing procedure is described in detail in Chapter 8. When we treat patients who have a history of anaphylaxis to a particular item, we use this method after completing the required NAET treatments and before the patient begins to use the item again.

If the patient was treated for a severe peanut allergy, (or severe reaction to milk, egg, wheat, fish, mushroom, etc.) after going through the required NAET treatments to neutralize the peanut, the patient is allowed to sit and hold the peanuts in a glass jar every day for 30 minutes for three days to a week. If the patient does not show any symptoms of previous allergy, he/she will be allowed to hold a peanut in the hand without a bottle for three to five days, 30 minutes daily. If that does not produce any allergic reaction, then the patient will be allowed to put a small piece of nut in the mouth and hold it there for five to ten minutes every day for a few days. If that also does not produce any reaction, the patient will be allowed to eat a small piece of the nut and observe the reaction. Usually, by this time, the patient will be able to use the allergen confidently without fear. Check with your practitioner for more details.

Commonly Used Standard Allergy Tests

Scratch Test

Western medical allergists generally depend on skin testing (scratch test, patch test, etc.), in which a very small amount of a suspected allergic substance is introduced into the person's skin through a scratch or an injection. The site of injection is observed for any reaction. If there is any reaction at the area of injection, the person is considered to be allergic to that substance. Each item has to be tested individually.

RADIO-ALLERGOSORBANT TEST (RAST)

The radio-allergosorbant test or RAST measures IgE antibodies in serum by radioimmunoassay and identifies specific allergens causing allergic reactions. In this test, a sample of patient's serum is exposed to a panel of allergen particle complexes on cellulose disks. Radiolabeled anti-IgE antibody is then added. This binds to the IgE-APC complexes. After centrifugation, the amount of radioactivity in the particular material is -directly proportional to the amount of IgE antibodies present. Test results are compared with control values and represent the patient's reactivity to a specific allergen. This test reveals immediate response to an allergen. Delayed reactions are not recorded in this test.

Blood counts, chemistry, lipid panels, etc. should be checked for any possible abnormalities. Usually autistic children show low levels of sodium in their body, and the autistic trait of craving salt.

Like Cures Like

Homeopaths believe that if an allergen is introduced to the patient in minute concentrations at various times, the patient can build up enough antibodies toward that particular antigen so that eventually, the patient's violent reactions to that particular substance may reduce in intensity. In some cases, reactions may subside completely and the patient can use or eat the item without any adverse reaction. Homeopaths believe that if an allergen is introduced to the patient in minute concentrations at various times, the patient can build up enough antibodies toward that particular antigen. Eventually, the patient's violent reactions to that particular substance may reduce in intensity. In some cases, reactions may subside completely and the patient can use or eat the item without any adverse reaction.

All of the above methods work on a certain percentage of people. It is for the practitioner to select the best tests suitable for the individual patient. All of the above methods work on a certain percentage of people. It is for the practitioner to select the best tests suitable for the individual patient.

4

Order of Treatments

Some people with environmental allergies suffer various levels of discomfort while they go through the 25-hour waiting period after an NAET treatment. If you get treated for Basic 15 first and strengthen your immune system, you will experience less post-treatment sickness. It is very important that you clear each treatment at 100% level.

If you desensitized each treatment completely with NAET at 100% level, then that allergen and associated ingredients of that allergen will not bother you with future contacts. These are the ingredients in the sample egg mix: egg white, egg yolk, chicken, feather, tetracycline. If the egg white is cleared and not egg yolk, products with egg yolk can still cause problem to the sensitive individual. If egg yolk is cleared and not the egg white, then products with egg white can cause problem with any future contact. Egg white is hidden in many products: Bread, pastries, candies, crackers, soups, salad dressings, hair shampoo, hair conditioner, skin creams, etc. can make the person uneasy or sick depending on the intensity of the sensitivity.

If the egg mix is not cleared at a 100% level, feather pillows can cause problem since egg mix contains energy of the feather. Chicken can also cause problem if the tetracycline is not cleared all the way. So after each NAET Basic-mix treatment, it is mandatory to test each individual ingredient to see if it is cleared. People with environmental or chemical sensitivities usually have problems clearing the Basic treatments on the first attempt because most people from this group become immune deficient from long-term suffering. Usually these people have approached various medical specialists and tried out many different treatments without much results by the time they discover they are suffering from environmental sensitivities. Most people from this group start out to be fairly healthy in their childhood or early adult days. Their subtle sensitivities from childhood usually go unnoticed. When they get slightly sick with different allergies, patients and doctors generally misdiagnose the conditions like "a common cold", or a "mild case of flu," "sinusitis," "indigestion," or "under the weather." Then later on, some incident takes places which can act like a trigger and a chain of health problems begins from there.

NAET BASIC FIFTEEN ALLERGENS
(See explanation below)

BBF
Egg mix
Calcium mix
Vitamin C mix
Vitamin B complex
Sugar mix
Iron mix
Vitamin A mix
Minerals mix (check water, drinking and tap)
Salt mix
Grain mix

Yeast mix
Stomach acid
Base (Digestive Enzymes)
Hormones
Immunizations/Vaccinations/ any drug taken in the past/
Drugs taking now.

PRIORITY

After the Basics your doctor will be able to treat you on a
priority basis, such as:

Daily samples collected from your food and drinks
Artificial sweeteners
Spice mix-1
Spice mix-2
Animal fat
Vegetable fat
Essential Fatty Acids
Dried beans
Amino acids
Vegetable proteins
Vegetable mix
RNA/DNA/CF
Gum mix
Gelatin
Alcohol
Food colors
Food additives
Vitamin E
Vitamin D
Vitamin F
Vitamin K

Vitamin T
Starch mix

Common Allergens Causing Environmental Disorders:

Acetate
Acrylic
Animal Dander
Animal Epithelial
Bacteria Mix (mycoplasma group)
Benzene
Braces & Retainers
Carpet
Carpet Glue
Carpet Lining
Ceiling
Ceramic Tiles
Ceramics
Chair
Chlorine
Coins
Cold
Cold mist
Copper
Cotton
Crude oil
Damp-heat
Dioxin
Door knob
Dry cleaning fluid
Dry wall
Dryness
Dust mix
Fabrics (cotton. polyester, acetate, etc).
Fiber glass

Flower Vase
Flowers
Foam
Food chemicals
Formaldehyde
Fungus
Furniture
General mail
Gold
Graphite
Grasses
Hair dryer
Halogen light
Heat
Heavy metal mix
Herbicides
High altitude
Humidity
Insects
Insulation
Jewelry
Latex
Marbles
MBT
Mercury
Microfungus
Mold
Neurotoxins
Newspaper
Newspaper ink
Nickel
Paint
Paper products (bills, money, coloring books)
Parasites
PBB (Poly Biphenyl Bromide)

Freedom From Environmental Sensitivities

PCB (Polychlorinated biphenyls)
PBDE (Polybromodiphenyl Ethers)
Pesticides
Phenolics
Physical agents
Plastic-hard
Plastic-soft
polish -wood
Pollens
Polyester
Pressed wood
Radiation (Sun, X-ray, Computer, TV, microwave)
Radon
Recycled paper
Remote controller
Roofing material
Silica
Silver
Smoking/nicotine
Sodium chloride
Swimming pool water
Sympathetic/ Parasympathetic nerves
Tap water
UV light
Vanadium
Virus mix - (EBV, hepatitis, influenza, blood, blood+virus)
Wall cabinet
Wall paint
Water chemicals
Water Filter Salt
Water pollutants
Weed-killer
Wheel chair
Walking stick
Weeds

Wood mix
Wool
Exercise
Allergy to people - (co-workers, family members, animals, pets, etc).

Supplements - Test and see what is needed. (calcium, B complex, sugar, minerals, essential fatty acids, liquid oxygen, co-enzyme Q 10, magnesium, hormone, adrenal support, thyroid support, lung support, liver support)

Any of these to keep the energy flowing properly through the energy pathways:

♦ Regular chiropractic manipulations, brain massage, body massage and massaging the self-balancing points at home twice daily from page 58 in *Living Pain Free* book by the author.

♦ Exercise - at least 10 minutes twice a day

♦ Emotional treatments as needed.

EXPLANATION OF NAET BASIC ALLERGENS

For 25 hours after treatment, you are allowed to eat only the foods listed herein:

1. BBF: (Brain-Body Balancing Factor.)

No avoidance of food.

2. Egg Mix: (egg white, egg yolk, chicken, tetracycline and feathers).

You may eat brown or white rice, pasta without eggs, vegetables, fruits, milk products, oils, beef, pork, fish, coffee, juice, soft drinks, water, and tea.

3. Milk and Calcium: (breast milk, cow's milk, goat's milk, casein, albumin, and calcium).

You may eat cooked rice, pasta, (no raw fruits or vegetables) cooked, heated and cooled fruit juices, and vegetables (like potato, squash, green beans, yams, cauliflower, sweet potato), chicken, red meat, salt, oils, and drink calcium-free water, coffee and tea without milk.

4. Vitamin C: (fruits, vegetables, vinegar, citrus, bioflavonoid, rutin, berry, vitamin C, ascorbic acid, Oxalic acid, and citric acid).

You may eat cooked white or brown rice, pasta without sauce, boiled or poached eggs, baked or broiled chicken, fish, red meat, brown toast, deep fried food, French fries, salt, oils, and drink coffee or water.

5. B complex vitamins

You may only eat cooked white rice, cookies and bread or pancakes made with white flour, cauliflower raw or cooked, well cooked or deep fried fish, salt, white sugar just for taste in a limited amount (no brown sugar or natural syrups), black coffee, French fries, and purified water when treating for any of the B vitamins. Rice should be washed well before cooking. It should be cooked in lots of water and drained well to remove the fortified vitamins. Since sugar is a helper of B vitamins, consuming too much sugar or eating sugary foods during the 25 hour period following B vitamins can cause treatment failure.

6. Sugar Mix: (cane sugar, corn sugar, maple sugar, grape sugar, rice sugar, brown sugar, beet sugar, fructose, molasses, honey, dextrose, glucose, and maltose).

You may eat white rice (white rice is pure starch, it is okay to eat since it takes time to convert into sugar inside the body; brown rice contains immediately available rice sugar in it, so it is not safe to eat), pasta without sauce, vegetables, vegetable oils, meats, eggs, chicken, water, coffee, and tea without milk. Avoid fruits.

7. Iron Mix: (animal and vegetable sources: beef, pork, lamb, raisin, date, seeds, nuts, dark green leafy vegetables, and broccoli).

You may eat white rice without iron fortification, sourdough bread without iron, cauliflower, white potato, chicken, light green vegetables (white cabbage, iceberg lettuce, white squash), yellow squash, and orange juice.

8. Vitamin A: (animal and vegetable source, beta carotene, fish and shell fish).

You may eat cooked white rice, products made from white flour, pasta, potato, cauliflower, red apples, chicken, water, and coffee.

9. Mineral Mix: (magnesium, manganese, phosphorus, selenium, zinc, copper, cobalt, chromium, gold, and fluoride).

Trace minerals (43 trace minerals)

You may use only distilled water for washing, drinking, and showering. You may not eat uncooked foods. You may eat cooked rice, vegetables, fruits, meats, eggs, salt, oils, and drink milk, coffee, tea and distilled water. Avoid root vegetables in any form - cooked or raw (onions, potato, turnip, carrot, etc.). All vegetables, fruits and grains contain trace minerals and macrominerals, but when you cook them, the naturally found organic minerals break down into their elemental form and they can be absorbed by human body without causing energy blockages. But root vegetables accumulate abundance of inorganic mineral from the soil where they grow and they store them under the skin. They do not break down or absorb as easily as the organic minerals and they can cause you treatment failure if you come in contact during the avoidance time.

10. Salt Mix: (sodium and sodium chloride, table salt, sea salt, rock salt, iodized salt, and water softener salts).

You may use distilled water for drinking and washing, cooked rice, cooked grains, fresh or cooked vegetables and fruits (except celery, and onions; avoid canned or preserved vegetables and vegetable products) meats, chicken, and sugar.

11. Corn Mix and **Grain mix:** (blue corn, yellow corn, cornstarch, corn silk, corn syrup). In some cases you may be able to treat corn and grains together at one time. Please check with your practitioner for the possibility of treating together. In severe allergies, they should be treated separately.

You may eat steamed or cooked vegetables, white rice, fish, chicken, and meat. You may drink milk, water, tea and/or coffee. Avoid cream or sugar (corn may be a fortified ingredient in them).

Grain Mix: (Wheat, rice, oat, millet, rye, gluten, corn, oats, millet, barley, kamut, cous cous, farina, and brown rice)

You may eat white rice, vegetables, potato, fruits, meats, milk, and drink water.

Avoid all products with gluten and grains.

11. Yeast Mix: (brewer's yeast, torula yeast, bakers yeast, candida, yogurt, whey).

Whey:

You may eat rice, vegetables, fruits, chicken, egg, turkey, beef, pork, beans, and lamb.

Yogurt:

You may eat vegetables, meat, chicken, and fish. No fruits, no sugar products. Drink distilled water.

12. Stomach acid: (Hydrochloric acid).

You may eat raw and steamed vegetables, cooked dried beans, eggs, oils, clarified butter, and milk.

13. Base: (digestive juice from the intestinal tract contains various digestive enzymes: amylase, protease, lipase, maltase, peptidase, bromelain, cellulase, sucrase, papain, lactase, gluco-amylase, and alpha galactosidase).

You may eat acid producing foods: sugars, starches, grains, breads, and meats.

Daily Food Samples Treatment: Start collecting a small portion of different food groups from every meal and treat for the mixture of breakfast, lunch and dinner. Collect this combined food sample daily and self-treat every night before bedtime.

14. Organ mix: (Bladder, Brain mix, Gall bladder, Heart, Kidney, Large Intestine, Liver, Lung, Ovary, Pancrease, Small Intestine, Spleen, Stomach, Uterus, Prostate)

15. Hormones: (estrogen, progesterone, testosterone)

You may eat vegetables, fruits, grains, chicken, and fish.

Source of Exposure: Eating or using red meats and products with hormones. If one is able to get the meat from an animal that has never received any hormone, it is OK to eat the red meat from that source. Avoid stimulating your own hormones. Avoid treating during menstrual period.

End of Basic-15. Now your practitioner may test and treat you on priority basis if it is indicated for your condition. Check with your NAET practitioner. After the priority is completed, you should continue through the list below to further improve your immune system and to prevent future reactions from other allergens.

16. Spice Mix 1: Spices are known to cause severe health problems like pain, fibromyalgia, arthritis, small and large joint pains, muscle aches, pins and needle sensation anywhere in the body, insomnia, chest pains, pain in the breast, intercostal neuralgia, peripheral neuralgia, vulvodynea, excessive sweating, constipation, etc. Spices are used in cooking in every part of the world. There are hundreds of spices used by people in cooking various dishes. Check individual spices and may treat them in group, but after cpmpleting the group allergy make sure each one is completed individually and in all possible combinations in order to get free from various pains.

In some cases you may treat spice 1 & 2 together. Check with your practitioner. (cardamon, cinnamon cloves, nutmeg, garlic, cumin, fennel, coriander, turmeric, saffron, and mint).

You may use all foods and products without these items.

17. Spice Mix 2: (peppers, red pepper, black pepper, green pepper, jalapeno, banana peppers, anise seed, basil, bay leaf, caraway seed, chervil, cream of tartar, dill, fenugreek, horseradish, mace, marjoram, MSG, mustard, onion, oregano, paprika, ginger, poppy seed, parsley, rosemary, sage, sumac, and vinegar).

You may eat or use all foods and food products without the above listed spices.

18. Artificial Sweeteners: (Sweet and Low, Equal, saccharine, Twin, and aspartame). There are many kinds of artificial sweeteners available today, which are refined chemical products and can irritate the brain and nervous system.

You may eat: anything without artificial sweeteners. Use freshly prepared items only.

19. Coffee Mix: (coffee, chocolate, caffeine, tannic acid, cocoa, cocoa butter, and carob). A caffeine- like substance is secreted by certain parts of the brain, and is very essential for the brain's normal function.

You may consume anything that has no coffee, caffeine, chocolate and/or carob.

20. Animal Fat: In some cases you may treat vegetable and animal fats together. Check with your practitioner. (butter, lard, chicken fat, beef fat, lamb fat, and fish oil).

You may use anything other than the above including vegetable oils.

21. Vegetable Fat: (corn oil, canola oil, peanut oil, linseed oil, sunflower oil, olive oil, palm oil, flax seed oil, and coconut oil).

You may use steamed vegetables, steamed rice, meats, eggs, chicken, butter, and animal fats.

22. Amino Acids: (essential amino acids: lysine, methionine, leucine, threonine, valine, tryptophane, isoleucine, and phenylalanine).

23. Amino Acids: (non essential amino acids: alanine, arginine, aspartic acid, carnitine citrulline, cysteine, glutamic acid, glycine, histidine, ornithine, proline, serine, taurine, and tyrosine).

You may eat cooked white rice, cauliflower, and iceberg lettuce.

24. Whiten All: (Food additive)

You may eat cooked vegetables, pasta, rice, meats, chicken, and eggs.

25. Turkey: (Serotonin)

You may eat any food that does not contain B1, B3, B6, tryptophane, and neurotransmitters (dopamine, epinephrine, norepinephrine, serotonin, acetylcholine).

26. Dried bean Mix: (vegetable proteins, soybean, and lecithin).

You may eat rice, pasta, vegetables, meats, eggs, and anything other than beans and bean products.

27. Food colors: (different food colors in many sources like: ice cream, candy, cookie, gums, drinks, spices, other foods, and/or lipsticks, etc.).

You may eat foods that are freshly prepared. Avoid carrots, natural spices, beets, berries, frozen green leafy vegetables like spinach.

28. Food additives: (sulfates, nitrates, phosphates, BHT).

You cannot eat hotdog or any prepackaged food. Eat anything made at home from scratch.

A list of other food additives used in prepackaged foods are given in detail at the end of this Chapter. Please check them and treat them if they apply to you. Many environmentally sensitive people react to these hidden additives making it difficult to buy or consume pre-made foods. Many of my patients prepared their own food before treating for these food chemicals. After they cleared the allergies to these additives, they enjoy their freedom to eat in restaurants like normal people.

29. Refined starches: (corn starch, potato starch, and modified starch). Refined starches are used as a thickening agent in sauces and drinks. Many people are allergic to starches. Refined starches should be avoided.

You may eat whole grains, vegetables, meats, chicken, and fish.

30. Baking powder/ Baking soda: (baked goods, toothpaste, and/or detergents).

You may eat or use anything that does not contain baking powder or baking soda including fresh fruits, vegetables, fats, meat, and chicken.

31. Night shade vegetables: Tomato, potato, eggplant, bell pepper and onion make up the nightshade vegetable group. Most people are missing the enzyme in their bodies that helps with the digestion of the special chemical contained in these vegetables. After NAET treatment, the body will turn on the function of producing these enzymes on demand and will begin to produce the special enzyme to digest these vegetables.

Avoid eating these vegetables for 25 hours after treatment.

32. Other Hormones: This is a good time to check the glands (thyroid, pituitary gland, ovary, adrenals, etc.) and their secretions.

33. Virus mix: (E.B.V., C.M.V., herpes simplex, herpes zoaster, influenza).

Source of Exposure: Contact with infected persons. If someone is infected with a virus, treat for the specific sample like herpes zoaster, etc. Also you may take a sample of your own body fluid (saliva, urine, stool, blood, skin tissue, etc.,) and treat for it.

Treatment Protocol: holding the saliva in a container self-balance the acupressure points (Figure 7-) at the first sign of cold or "flu" and you will be relieved of the symptoms in just a few minutes and you may not get the "flu".

Avoid contact with people who have any infections. Avoid uncooked foods, old and stale food. Eat freshly cooked food and drink freshly boiled cooled water. Distilled water is not sufficient to replace freshly boiled cooled water.

34. Bacteria Mix

(Staphylococcus aureus, streptococcus (viridans & non-hemolytic), streptococcus, pneumoniae & klebsiella pneumoniae).

Source of Exposure: Contact with infected surfaces.

Avoid contact with people who have any infections. Avoid uncooked foods, old and stale food. Eat freshly cooked food and drink freshly boiled cooled water. Distilled water is not always

safe. People have developed sickness from distilled water bought
in sealed bottles.

35. Parasites

Source of Exposure: Uncooked food, vegetables, tap water
and contaminated water. Many people host various parasites in
the body. They find their way into our body via uncooked vegetables, meats, fish and unwashed fruits. Once they get in, they
multiply rapidly and find their way into the blood stream. Parasite
infestation can cause various health disorders in the body, including upper respiratory infections, asthma, brain fatigue, brain fog,
fibromyalgia, insomnia, general itching, abdominal bloating, pain in
the abdomen, sinusitis, diarrhea, anal itching, hyperactivity, irritability, mood swings, and unexplained weight gain.

36. Water: (drinking water, tap water, filtered water, city
water, lake water, rain water, ocean water, and river water).

MCS people can react to any water, by drinking or touching.
Treat them as needed and avoid the item treated for 25 hours.

37. Chemicals: (chlorine, swimming pool water, detergents,
fabric softeners, soaps, cleaning products, shampoos, lipsticks, and
cosmetics you or other family members use, Clorox, bleach, chemical fumes from nearby factories, auto shops, etc.). Some of the
commonly seen chemicals are listed below. They are used in different combinations in the products.

(Acetonum, acid acetic, acid benzoic, acid citricum, acid
glutaminicum, acid malicum, acid oxaceticum, acid pyruvicum, acid
sorbicum, acid uricum, acrylic resins, aflatoxin, ammonium
benzoicum, ammonium carob, ammonium idatum, ammonium
phoshoricum, ammonium valerianicum, amyl nitrosum,
anthracenum, antimonium crudum, antimonium tartaricum,
acetylcholinchloride, acid asparagine, acid cisacontitum, acid
fumaricum, acid hydrochloric, acid nitricum, acid phosphoric, acid
saliculium, acid succininum, acrylate, adipinic acid, aluminum, am-

monium bromide, ammonium causticum, ammonium muriaticum, ammonium picratum, amyl alcohol, aniline, antimonium Arsenicosum, ant. sulphuratum auratum, argentum metallicum, argentum nitricum, asbestos, autocrylate, bar iodide, benzene, benzochinon, borax, perchloroethylene, phenylendiamine, phosphate cement, polypeptide aga, propylthiouracil, silica, sorbic acid, sulfaurea, toluol, urethanum, xylol, zincum oxidatum, zincum valeriante, arsenic, aurum metallicum, barbitonic acid, bar oxalsuccinicum, benzoic acid, bismuth.met.,petonum,petroeum, phenylmercurinitrate, picric acid, polyvinyl, pyruvic acid, sincore, succinic acid, sulphur, urea, vinblastinsulfate, zinblastinsulfate, zincum cyanatum, zincum picratum).

Source of Exposure: Contact with the above items. Wash your clothes in plain water prior to treatment.

Ask the patient to collect samples of chemicals from local area and treat it while smelling on next visit.

Avoid the above items during and for 25 hours after treatment.

Given below are environmental allergens and chemical compounds. You may treat them on a priority basis after Basic fifteen.

38. Inhalants:

Avoid pollens, weeds, grasses, flowers, wood mix, room air, outside air, smog, and polluted air from nearby factories.Most of thse inhalants produce similar health disorders in people. Some may just suffer from one problem, and others may siffer from many from this following list. Check out the list and mark your symptoms before you see the NAET practitioner for treatment for inhalants.

Wear a mask while going outdoors after treatment.

39. Grasses

Source of Exposure: Alfalfa, Barley-cultivated, Bermuda grass, Blue-Canada, Bluegrass, Brome-Hungarian, Canary grass, Clover-sweet, Corn-pollen, Grama-blue, Johnson grass, Meadow Fesue, Oat-cultivated, Quack grass, Rye, Rye grass, Rye grass-perennial, Sorghum grain, Sweet Vernal grass, Velvet grass, Wheat grass-Western, Bahia, Bent-creeping, Blue-annual, Blue-Kentucky, Brome grass, Chess-Southern, Corn-cultivated, Five Grass mix, grass pollen, June grass-Western, Oat grass-Western, Oat Grass-tall, Orchard Grass, Redtop, Rye-cultivated, Rye Grass-Italian, Salt Grass, Sudan Grass, Timothy Grass, Wheat-cultivated. Walking on the lawn, going outdoors. Wear shoes and socks while walking outside

Treatment protocol: First day treat the NAET sample of grass. On the following visit, if the sample is passed, then the smell of cut grass should be treated. Ask the patient to bring a fresh sample. Patient only smells during the NAET not while waiting for 20 minutes. Wear a mask while going outdoors after treatment.

40. Pollen Mix

Source of exposure: Going outdoors

Keep a pan of water in a windy area for four to six hours or more. Collect the water and treat for the local pollens along with the NAET sample.Wear a mask while going outdoors after treatment.

41. Weed Mix

(Broom Weed, Cocklebur, Dock-sour, Allscale, Careless Weed, Dock-Rumex Mix, Firebush, Goldenrod, Hemp-common, Lambsquarter, Marsh Elder-narrow leaf, Mexican Tea, Pigweed-rough, Pigweed mix, Rabbit Bush, Ragweed-false, Ragweed-short, Ragweed- Southern, Ragweed- Western Giant, Ragweed Mix, Sagebrush-common, Sage-pasture, Sage Mix, Shadcale, Sugar Beet Pollen, Western Water Hemp, Winterfat, Wormwood-Annual, Dock-yellow, Greasewood, Jerusalem Oak, Marsh Elder-Burweed,

Marsh Elder-rough,Mugwort-common, Pigweed-spiny, Plantain-English, Quail Bush, Ragweed-desert, Ragweed-giant, Ragweed-slender, Ragweed-Western, Ragweed-Woolly, Russian Thistle, Sage-Dragon, Sage-Prairie, Saltbush-annual, Sheep Sorrell, Weed Pollen Mix, Wingscale, Wormwood-common).

Source of Exposure: Going outdoors. Wear a mask while going outdoors after treatment.

First day, treat the NAET sample. Ask the patient to collect samples of weeds from local area and treat it on next visit. Wear a mask while going outdoors after treatment.

42. Wood Mix

Source of Exposure: Contact with woods, things made with woods, pressed wood, furniture, wall cabinet, bed frame, side tables, wood cabinets in the kitchen, bathrooms, book shelves, wood polishes, walking stick, roof, ceilings, stairways, floor, dining tables, chairs, desk, doors, You may wear a pair of gloves to avoid contacts with wooden surfaces.

43. Flower mix.

Source of Exposure: Flowers and perfumes and going outdoors.

Wear shoes and socks while walking outside. Wear mask and gloves if you are going out. Do not smell perfume.

TXP: First day treat this sample. On next visit, if he/she passed the sample, treat him/her for the smell of the flowers. Ask the patient to bring real live flowers and make the patient smell while you treat the spinal points. Patient does not have to smell after the spinal treatment while getting needles or gate treatments.

Avoid contact with all flowers, and perfumes. wear a mask for 25 hours after treatment.

44. Tree Mix

Source of Exposure: (Acacia, Alder-red, Alder-white, Ash-Arizona, Ash-green, Ash-Oregon, Aspen-Quaking, Bayberry, Beech-American, Beefwoood, Birch-mixed, Birch-river, Box-Elder, Cedar-Pinchot, Douglas fir, Eucalyptus, Birch-red, Birch-white, Cedar-mountain, Cedar-red, Cottonwood-Arizona, Cottonwood-Eastern, Cottonwood Mix, Cottonwood-Western, Cypress-Bald, Elm-American, Elm-Chinese, Elm-Cedar, Elm Mix, Elm-slippery, Gum-black, Gum-sweet, Hackberry, Hazelberry, Hazelnut-American, Hemlock-Western, Hickory-mixed, Hickory-Pignut, Hickory-Shellbark, Hickory-white, Iodone Bush, Juniper-one-seeded, Juniper-Rocky Mountain, Magnolia, Maple-hard, Maple-soft, Mesquite, Mulberry-red, Mulberry-paper, Mulberry-white, Oak-black, Oak-Blackjack, Oak-live, Oak-bur, Oak-post, Oak-red, Oak-white, Olive-European, Osage-orange, Palm-date, Pecan, Pepper, Pine-mixed, Poplar-Lombardy, Poplar-white, Privet, Spruce-blue, Sumac-Upland, Sycamore, Tree of Heaven, Walnut-black, Willow-black, Willow-Pussy.)

Source of Exposure: Going outdoors, and going near trees.

Wear shoes and socks while walking outside. Wear mask and gloves if you are going out.

45. Mold Mix

Source of Exposure: Dirty and moldy house, living near backwater bay, ocean, pools, canals, etc.. Avoid treating on a cloudy, rainy day Clean up the house well. Keep the house dry. Stay away from leaky bathrooms, old houses etc. Wear freshly washed clothes during treatment.

Ask the patient to collect samples of molds from local area and treat it on the following visit.

46. Fabric Mix

Try to treat one kind of fabric first, like cotton or polyester, etc. Then wear the allergy-cleared item while treating for the fabric mix.

Source of Exposure: Cotton, foam, rayon, sheep wool, spandex, orlon, polyester, silk, fleece, acrylic, and nylon, carpet, carpet glue, carpet lining,. contact with the fabric that is being treated.

47. Formaldehyde

Source of Exposure: New buildings, new clothes, newspaper, liquid paper, pressed woods, paints, paint thinner, fumes, perfumes, certain ice creams. Wear a mask and use a pair of gloves. Remove name tags from the clothes or tape them with masking tape.

Wear mask and gloves if necessary

TXP: First day treat the energy of the sample. On next visit, if he/she passed the sample, treat him/her for the smell of the formaldehyde (if the patient is highly sensitive) while you do NAET. Treat for fabrics (daily and sleep attire; towels, bed linens, blankets, formaldehyde).

Treat each kind of fabric separately and avoid the particular cloth or kind of cloth for 25 hours.

48. Latex products : (shoe, sole of the shoe, elastic, rubber bands, and/or rubber bathtub toys, computer parts, certain household items, certain cosmetic products, elastics on various products, clothes, work materials, etc.).

Avoid latex products until NAET treatments are completed successfully.

49. Plastics

Source of Exposure:- All plastic and crude oil products including computer key boards, pens, vinyl chairs, Plastic-hard, Plastic-soft, containers, book covers, toothbrush, hair brush, toys, play or work materials, utensils, toiletries, plastic portion of wheel chair, walking stick, car, Remote controller, cell phone, and/or phone etc. Other chemicals associated with plastics, computers, radiation, etc., seen in the environments and cause many health problems are due to:

PBB (poly biphenyl bromide)

PCB (Polychlorinated biphenyls)

PBDE (Polybromodiphenyl Ethers)

Avoid contact with products made from plastics. Wear a pair of cotton gloves.

50. Crude oil/Synthetic materials

Source of Exposure: Gasoline, plastic products, latex products, school work materials (crayons, coloring paper and books, inks, pencils, glue, play dough, other arts, and craft materials.

Avoid using them or contacting them during and for 25 hours after treatments. Wear a pair of gloves if you have to go near them.

51. Smoking: Nicotine, Tobacco

Source of Exposure: Smoking areas, smoke from cigarettes, clothes and substances making contact with cigarette smoke, banana, malt, cow's milk, potato, tomato and yeast mix. You may wear a mask for 25 hours.

TXP: First day treat this sample. On the following visit, after passing the first treatment, treat him/her for the smell of the smoke. Place wet paper towelor cotton balls in a glass jar with a lid. Ask the patient to fill up the glass jar with cigarette smoke (or whatever he/she is smoking). Then make the patient smell the smoke while giving NAET. Patient does not have to smell after the initial NAET while waiting for 20 minutes.

Avoid primary and secondary smoke. Avoid wood burning smoke too. Wear a mask.

52. Dust Mix and dust mites.

Source of Exposure: Dusty areas.

Clean up the living area before the treatment. Wear a mask for 25 hours.

Ask the patient to collect samples of dust from his/her house or from local area and treat it on next visit.

Avoid dust during and for 25 hours after treatments.

53. Heavy metals: (Mercury mix**,** amalgam, lead, cadmium, aluminum, arsenic, copper, gold, silver, silica, silver and vanadium. Check braces & retainers).

Source of Exposure: Fish and fish products, touching your mouth if you have dental fillings, pesticides.

Commonly Seen Allergic Symptoms: Abdominal bloating, flatulence, body aches, insomnia, asthma, bronchitis, depression, digestive disorders, depletion of Zinc, emotional imbalances, fatigue, flu-like symptoms, frequent colds, headaches, heart palpitation, high blood pressure, hyperactivity, irritability, joint pains, muscle pain, poor concentration, skin disorders, respiratory disorders, sinusitis, shortness of breath, skin problems, sleep disorders, upper respiratory disorders, water retention, weakness of the muscles, and weak limbs

You may use only distilled water for drinking, washing and showering. You may eat only cooked rice, vegetables, fruits, meats, eggs, milk, coffee, and tea.

Wear a pair of gloves for 25 hours after treatment.

54. MSG: (monosodium glutamate).

Source of Exposure: Normal Chinese food, and anything prepared with MSG or Accent.

Treatment Protocol: Treat the 15 NAET basics and spice mix 1 and 2, amino acids, food additives, whiten-all, food colors, before you try to treat MSG on a severely allergic person. Many people are highly sensitive to MSG causing circulatory, respiratory and cardiac problems and even anaphylaxis. If you have a history of severe reactions like anaphylaxis, you should inform your NAET practitioner, so that he/she could use the special anaphylaxis protocol on you while you get treated with NAET.

You may eat freshly prepared vegetables, fruits, meat, and grains without MSG.

55. Paper Products, Newspaper/ Newspaper Ink:
(newspaper, paper goods, milk cartons, paper bills, paper money, recycled paper, ink, reading books, coloring books, with colored illustrations)

Source of Exposure: Touching paper goods, newspaper, facial paper, hand towel, pine products, tissue paper etc.

Ask the patient to collect samples of newspaper from local area and treat it on next visit.

56. Perfume

Source of Exposure: Perfumed soaps, cosmetics, hair sprays, room deodorizers, soaps, flowers, perfumes, skunk oil, after-shave, flowers.

Wear shoes and socks while walking outside. Wear mask and gloves if you are going out. Do not smell perfume.

TXP: First day treat this sample. On next visit, if he/she passed the sample, treat him/her for the smell of the flowers. Ask the patient to bring real live flowers and make the patient smell while you treat the spinal points. Patient does not have to smell after the spinal treatment while getting needles or gate treatments.

Avoid perfume and any fragrance from flowers or products containing perfume for 25 hours after treatment.

57. Pesticides

(Antikeimmetel A, cyol hahm, diphenylamine, dichlorvos (DDVP), HCC-B (endosulfan intermed.), naphthalene HCL, pentachlorphenol, dorphosina, paraquat, aminotrazol, heptachlor, atrazine, aldicarb, methyl Mercaptan, sodium pyrophosphoric, parathion, para dichlorobenzene, superphosphate, calicum cyanide, DDT, isopropyl-N phenylcarbamate, HCC (lindane), dinitrocresol, diazonine, trichphim, 2,4,5 tester, toxaphene, hexachlorbenzol, Endosulphan, dithiocarbamate (ferbam) 2-mercaptobenzothiazol, sodium-o-phenylpholate, sodium sulfate, polychlorinated biphenyl, phthalate B Ester, thiomasmerol.)

Source of Exposure: Fresh vegetables, fruits, meats, insecticides, new mattress, malathion sprays, ant baits, and house, grass, weeds, lawns, trees that have been sprayed for pesticides.

Avoid meats, grasses, ant sprays, and pesticides during and for 25 hours after treatment.

58. Insect mix: in infancy or childhood (bee, ant, wasp, spider, flea, or cockroach, fire ants, etc.).

Source of Exposure: Touching or going near any insects

Treat for the individual insect and avoid it while treating.

59. Radiation: (computer, television, microwave, X-ray, UV light, PBDE (Polybromodiphenyl ethers, PCB (Polychlorinated biphenyls and the sun radiation).

Source of Exposure: Sun, T.V., microwave, X-ray & computers.

Make the patient touch the TV or Computer Screen while the Computer is on and treat the spinal points if the patient is reacting to the radiation from the screen.

60. Drugs: given in infancy, during childhood or taken by the mother during pregnancy (antibiotics, analgesics, antidepressants, sedatives, laxatives, or recreational drugs).

Avoid the drug for 25 hours after treatment.

61. Immunizations: and vaccinations either you received or your parent received before you were born (DPT, PO-LIO, MMR, small pox, chicken pox, influenza, or hepatitis).

Nothing to avoid except infected persons or recently inoculated persons if there are any near you.

The items listed below are treated as needed and on a priority based protocol, which your NAET practitioner will explain to you.

62. Histamine

Whenever there is an allergic reaction in the body, special cells (mast cells) release histamine. You can be allergic to your own histamine. When that happens, histamine is produced very frequently in the body. It doesn't stop until the mechanism is turned off. Another way to turn off histamine is to take antihistamine, either in a medication or a natural way with vitamin C (that is if you are not allergic to it). It is very easy to treat for your own histamine with NAET, then your body will adjust to your histamine by itself to a normal level.

63. Neurotransmitters : (dopamine, epinephrine, norepinephrine, serotonin, acetylcholine).

You may eat anything other than milk products, and turkey.

64. Allergies to animals: Animal Epithelial/Animal Dander and pets;

Source of Exposure: Contact with the animals, their saliva, hair, danders, any other products made from animals or used by the animals. If you have a pet, make arrangements to stay away from the pet for 25 hours.

TXP: First day treat this sample. On next visit, if he/she passed the sample, treat him/her for the smell of the animal hair while you treat the spinal point. Patient does not have to smell after the spinal treatment.

Avoid the treated ones for 25 hours.

65. Freon

Source of Exposure: Air conditioned areas, soft plastic products (fluoromethane).

66. Allergies to people, (mother, father, caretakers, doctors, teachers, co-workers) Avoid the treated person for 25 hours.

67. Tissues and secretions: (DNA, RNA, thyroid hormone, pituitary hormone, pineal gland, hypothalamus, or brain tissue, liver, blood, and saliva).

68. Urine Sample

Take a sample of the morning urine and treat once a week until you finish treating all your allergies. Teach the patient how to do it. This helps to keep your allergies under control.

69. Emotional allergies: (fear, fright, frustration, anger, low self-esteem, and/or rejection, etc.). Fear is a major factor for environmentally ill people. With ongoing aches, pain, and other discomfort for years, their emotions have been hurt many times. They are afraid to do anything for the fear of pain and discomfort. They are afraid to eat, wear normal clothes, use normal cleansing or cosmetic products, go places, meet strangers, afraid of crowds, people, weather changes, etc. Probably childhood traumas may be found in this group of people too. Fear for each of these factors should be checked seperately and treated if necessary.

Nothing to avoid for emotional treatment.

After clearing the allergy to nutrients, appropriate supplementation with vitamins, minerals, and enzymes etc., is necessary to make up the deficiency and promote healing. Please read the guidebook for information on how to take supplement correctly.

70. Carob, a staple in many health food products, is another item that causes brain irritability among allergic people. Many health-conscious people are turning to natural food products in which carob is used as a chocolate and cocoa substitute. It is also used as a natural coloring or stiffening agent in soft drinks, cheeses, sauces, etc. We discovered that some of the causes of "holiday flu" and suicide attempts are allergies to carob, chocolate, and turkey.

If you have allergies you must look for these ingredients and additives in the food products you buy from the market.

Please read the labels. Manufacturers usually list these items on the cover of the product-container.

Aldicarb: It is an organic chemical water pollutant, seen often in city water. When the concentration of this chemical gets high in the city water, many people get sick with gastrointestinal disorders, like nausea, vomiting, pain, bloating, stomach flu, etc. Boiling the water for 30 minutes could help reduce the reaction. If your child is not allergic to apple cider vinegar, adding two-three drops of vinegar in eight ounces of water might help. **Alginates** (Alginic acid, algin gum, ammonium, calcium, potassium, and sodium alginates, propylene glycol alginates): Most of these are natural extracts of seaweed and used in the food industry primarily as stabilizing agents.

Propylene glycol is an antifreeze. This is supposed to be a safe solvent, used in food preparation, especially in ice creams. Alginates help to retain water. It helps to prevent ice crystal formation; helps uniform distribution of flavors through foods. They add smoothness and texture to the products and are used in ice creams, custards, chocolates, chocolate milk, cheese, salad dressings, jellies, confections, cakes, icings, jams, and some beverages.

Acetic Acid (sodium acetate and sodium diacetate). This is a common food additive. This is the acid of vinegar. Acetic acid is used as an acidic flavoring agent for pickles, sauces, catsup, mayonnaise, wine, foods that are preserved in vinegar, some soft drinks, processed cheese, baked goods, cheese spreads, sweet and sour drinks and soups. It is also naturally found in apples, cocoa, coffee, wine, cheese, grapes, and other over-ripened fruits. If your child gets allergic reaction to these natural foods he/she may be allergic to acetic acid.

Agar (Seaweed extract): This is a polysaccharide that comes from several varieties of algae and it can turn like a gel if you dissolve it in water. So this is used in ice cream, jellies, preserves, icings, laxatives, used as a thickening agent in milk, cream, and used as gelatin (vegetable form). This is a safe additive, but if your child is allergic to seafoods you may need to eliminate the allergy for this.

Albumin (cow milk-albumin): Many children are allergic to albumin in the milk. Researchers have found children/people who are allergic to milk albumin are at high risk to get any of these disorders: ADD, ADHD, Autism, bipolar diseases, schizophrenia, and other allergy-related brain disorders. NAET can desensitize you for milk-albumin.

Aluminum Salts (alum hydroxide, alum potassium sulfate, sodium alum phosphate, alum ammonium sulfate, and alum calcium silicate).: Aluminum salts are used as a buffer in various products. This helps to balance the acidity. Used as an astringent to keep canned produce firm, to lighten food texture, and used as an anti-caking agent.

Sodium aluminum phosphate is used in baking powder and in self-rising flours. Alum is used as a clarifier for sugar and as a hardening agent. Aluminum hydroxide is used as a leavening agent in baked goods. It is a strong alkali agent that can be toxic but when used in small amounts it is fairly safe. It is also used in antiperspirants and antacids. Aluminum ammonium sulfate is used as an astringent, and neutralizing agent in baking powder and cereals. It can cause burning sensation to the mucous membranes. Overuse of aluminum products may lead to aluminum toxicity and it can affect the brain chemistry. Other sources of aluminum are cookware, deodorants, antacids, aluminum foils, cans and containers.

Benzoates (sodium benzoate).: Benzoic acid occurs naturally in anise, berries, black olive, blueberries, broccoli, cauliflower, cherry bark, cinnamon, cloves, cranberries, ginger, green grapes, green peas, licorice, plums, prunes, spinach, and tea. Benzoic acid or sodium benzoate is commonly used as a preservative in food processing. This is used as a flavoring agent in chocolate, orange, lemon, nut, and other flavors in beverages, baked products, candies, ice creams, and chewing gums and also used as a preservative in soft drinks, margarine, jellies, juices, pickles, and condiments.

This is also used in perfumes and cosmetics to prevent spoilage by microorganisms. Benzoic acid is a mild antifungal agent. It is metabolized by the liver. Large amount of benzoic acid or benzoates can cause intestinal disturbances, can irritate the eyes, skin, and mucous membranes. This causes eczema, acne and other skin conditions in sensitive people.

Cal. Proprionate: (sodium proprionate and proprionic acid): These are found in dairy products, cheese, breads, cakes, baked goods and chocolate products, They are used as preservatives and mold inhibitors. They reduce the growth of molds and some bacteria.

Source: Baked products, breads, rolls, cakes, cup cakes, processed cheese, chocolate products, preserves, jellies, and butter.

Cal. Silicate: Used as an anticaking agent in products, table salt and other foods preserved in powder form used as a moisture control agent.

Carbamates: These pesticides are used widely in many places. Their toxicity is slightly lesser than some other pesticides like organochlorines. They are known to produce birth defects.

Source: pesticide-sprayed foods.

Carbon Monoxide: CO is an odorless, colorless gas that competes with oxygen for hemoglobin. The affinity of CO for hemoglobin is more than 200-fold greater than that of oxygen. CO causes tissue hypoxia. Headache is one of the first symptoms, followed by confusion, decreased visual acuity, tachycardia, syncope, metabolic acidosis, retinal hemorrhage, coma, convulsions, and death.

Source: Driving through heavy traffic, damaged gas range, leaky valves of the gas line, exhaust pipes, living in a closed up room for long time, trapped firewood smoke, smoke inhalation from being in a closed, running car, an automobile kept running in closed garage for hours, exhaust from autos and other machinery, etc.

Casein: Milk protein. Also used in prepared foods, candies, protein shakes, etc.

EDTA: This is a very efficient polydentate chelator of many divalent or trivalent cations including calcium. This is used primarily in lead poisoning. This is toxic to the kidneys. Adequate hydration is necessary when you take this in any form.

Ethylene gas (used on fruits, especially on green bananas).

Food Bleach: Most of these are used in bleaching the flour products. Benzoil peroxides, chlorine dioxides, nitrosyl chlorides, potassium bromate, mineral salts, potassium iodate, ammonium sulfate, ammonium phosphate, are the most commonly used food bleaches. They whiten the flour. They also improve the appearance. Whatever they are using should be listed on the labels. Sometimes more than one item is used for better benefit.

Formic Acid: This is a caustic, colorless, forming liquid. Naturally seen in ants (ant bite), synthetically produced and used in tanning and dyeing solutions, fumigants and insecticides. This is also used as an artificial flavoring in food preparations.

Malic Acid: A colorless, highly water soluble, crystalline substance, having a pleasant sour taste, and found in apples, grapes, rhubarb, and cactus. This substance is found to be very effective in reducing general body aches. If you are allergic to it, then you can get severe body ache.

Mannan: Polysaccharides of mannose, found in various legumes and in nuts. Allergy to this factor in dried beans causes fibromyalgia-like symptoms in sensitive people.

Mannitol: It is hexahydric alcohol, used in renal function testing to measure glomerular filtration. Used intravenously as an osmotic diuretic.

Salicylic Acid: Amyl, phenyl, benzyl, and methyl salicylates).

A number of foods including almonds, apples, apricots, berries, plums, cloves, cucumbers, prunes, raisins, tomatoes, and wintergreen.

Salicylic acid made synthetically by heating phenol with carbon dioxide is the basis of acetyl salicylic acid. Salicylates are also used in a variety of flavorings such as strawberry, root beer, spice, sarsaparilla, walnut, grapes, and mint.

Succinic Acid: Found in meats, cheese, fungi, and many vegetables with its distinct tart, acid taste.

Source: asparagus, broccoli, beets, and rhubarb.

Talc (magnesium silicate): Talc is a silica chalk that is used in coating, polishing rice and as an anticaking agent. It is used externally on the body surface to dry the area. Talc is thought to be carcinogenic. It may contain asbestos particles. White rice is polished and coated with it.

Tartaric Acid: This is a flavor enhancer. It is a stabilizer.

Commonly seen Water Chemicals (in drinking water).

Alum sulfate, ammonium chloride, benzene, carbon tetrachloride, chlorine, DDT, ferric chloride, gasoline, heavy metals (mercury, silver, zinc, arsenic, lead, copper), organochlorides, organophosphates, PCBs, pesticides, petroleum products, Sodium hydroxide, toluene, and xylene.

Commonly seen Water pollutants: There are many water pollutants we see in our water. Some of them get filtered out by the time we receive in our tap. Most of these pollutants still remain in small amounts. Some of these are inorganic water pollutants like: arsenic, asbestos, cadmium, chromium, copper, cyanide, Lead, mercury, nickels, nitrates, nitrosamines, selenium, silica, silver, and zinc.

Organic chemical water pollutants: 1,2, dichloroethane, 2,4,5,T, 2,4,-D., aldicarb, benzene, carbon tetrachloride, chloroform, DDT, dibromo-chloropropane (DBCP), dichlorobenzene, dioxane, endrin, ethylene dibromide (EDB), gasoline, lindane, methoxychlor, polychlorinated biphenyls (PCB), polynuclear aromatic hydrocarbon (PAH), tetrachloroethylene, toluene, toxaphene, trichloromethane, trichloroethylene (TCE), vinyl chloride, MTBE (Methyl tertiary butyl ether is a gasoline additive), and xylene.

Atrazine: weed killer known to pollute drinking water. A recent study reported that male reproduction studies around Missouri and Iowa farm country vs. controls show an atrazine impact link. Any pesticide can cause nerve damage, neurological disorders and infertility among other things.

Some people with extreme sensitivity to these pollutants react badly with exposure by exhibiting mild to severe symptoms in various health areas. Some of the commonly seen symptoms are nausea, vomiting, diarrhea, abdominal cramps, brain fog, fatigue, body ache, joint pains, water retention in the body, flu like symptoms, fever, eczema, and rashes, sinusitis, post nasal drips, insomnia, etc. If you boil it for 30 minutes, the effect of these chemicals is reduced.

When assessing a patient with sensitivity, care must be taken not to misdiagnose him/her.

5

Symptoms of Meridians

T he human body is made up of bones, flesh, nerves and blood vessels, which can only function in the presenceof vital energy. Like electricity, vital energy is not visible to the human eye.

No one knows how the vital energy gets into the body or how, when or where it goes when it leaves. It is true, however, that without it, none of the body functions can take place. When the human body is alive, vital energy flows freely through the energy pathways. Uninterrupted circulation of the vital energy flowing through the energy pathways is necessary to keep the person alive. This circulation of energy makes all the body functions possible. The circulation of the vital energy makes the blood travel through the blood vessels, helping to distribute appropriate nutrients to various parts of the body for its growth, development, functions, and for repair of wear and tear.

NAET has its origin in Oriental medicine. But if one explores most Oriental medical books–acupuncture textbooks, one may not find the NAET interpretation of health problems that I write in my books, because NAET is my sole development after observing my own reactions, my family's and patients' over the past two decades. Recently, however, Information about the effectiveness of NAET has been given credit in a number of books, but the reader will find correct information of NAET interpretation of Oriental medical principles only in my books.

In this book, information about acupuncture meridians are kept to a minimum, enough to educate the reader about some traditional functions and dysfunctions of the meridians in the presence of energy disturbances. Some of this information is also available in acupuncture textbooks that one may find in libraries. It is a good idea to have some understanding of Oriental medicine and the meridians when undergoing NAET treatments although it is not mandatory. To learn more about acupuncture meridians and mind-body connections, please read Chapter 10 in my book, *Say Good-bye to Illness.*

NAET utilizes a variety of standard medical procedures to diagnose and then treat allergies and allergy-related health conditions. These include: standard medical diagnostic procedures and standard allergy testing procedures (read Chapter 3) and an electrodermal computerized allergy testing machine to detect allergies. After detecting allergies, NAET uses standard chiropractic and acupuncture/acupressure treatments to eliminate them. Various studies have proven that NAET is capable of erasing the previously encoded incorrect message about an allergen and replacing it with a harmless or useful message by reprogramming the brain. This is accomplished by bringing the body into a state of "homeostasis" using various NAET energy balancing techniques.

Chiropractic theory postulates that a pinching of the spinal nerve root(s) may cause nerve energy disturbance in the energy pathways causing poor nerve energy supply to target organs. When the particular nerve fails to supply adequate amounts of energy to the organs and tissues, normal functions and appropriate enzymatic functions do not take place. The affected organs and tissues then begin to manifest impaired functions in digestion, absorption, assimilation and elimination. An allergy can also cause impaired functions of the organs and tissues themselves. In chiropractic theory, an allergy can be seen as a result of a pinched nerve. Impaired functions of the organs and tissues will improve when the pinching of the spinal nerves is removed and energy circulation is restored.

Oriental medical theory explains the same theory from a different perspective. In Oriental medicine, the Yin-Yang state represents the perfect balance of energies (the state of homeostasis). Any interference in the energy flow or an energy disturbance can cause an imbalance in the Yin-Yang state and an imbalance in "homeostasis." Any substance that is capable of creating an energy disturbance in one's body is called an allergen. The result of this energy disturbance is called an allergy.

According to NAET theory, when a substance is brought into the electromagnetic field of a person, an attraction or repulsion takes place between the energy of the person and the substance.

ATTRACTION

If two energies are attracted to each other, both energies benefit each other. The person can benefit from association with the other substance. The energy of the substance will combine with the energy of the person and enhance functional ability. For example: After taking an antibiotic, the bacterial infection is diminished. Here the energy of the antibiotic joins forces with the energy of the body and helps to eliminate the bacteria. Another example is taking vitamin supplements (if one is not allergic to them) and the gaining of energy and vitality.

REPULSION

If two energies repel each other, they are not good for each other. The person can experience the repulsion of his/her energy from the other as a pain or discomfort in the body. The energy of the person will cause energy blockages in his/her energy meridians to prevent invasion of the adverse energy into his/her energy field. For example: After taking a repelling antibiotic, not only does the bacterial infection not get better but the person might break out in a rash all over the body causing fever, nausea, excessive perspiration, light-headedness, etc. Another example is taking a repelling vitamin supplement one night and waking up with multiple joint pains and general body-ache the next morning. If repulsion takes place between two energies, then the substance that is capable of producing the repulsion in a living person is considered an allergen. When the allergen produces a repulsion of energy in the electro-magnetic field, certain energy disturbance takes place in the body. The energy disturbance caused from the repulsion of the substance is capable of producing various unpleasant or adverse reactions. These reactions are considered "allergic reactions."

IMMUNOGLOBULINS

In certain instances, the body also produces many defensive forces like "histamine, immunoglobulins, etc." to help the body overcome the unpleasant reactions from the interaction with the allergen. The most common immunoglobulin produced during a reaction is called IgE (immunoglobulin E). These reactions are called IgE-mediated reactions. During certain reactions, specific immunoglobulins are not produced. These are called non-IgE mediated reactions. Different types of immunoglobulins are produced during allergic reactions.

An allergy means an altered reactivity. Reactions and after–effects can be measured using various standard medical diagnostic tests. Energy medicine has also developed various devices to measure the reactions. Oriental medicine has used "Medical I Ching" since 3,322 BC. Another simple way to test one's body is through simple, kinesiological neuromuscular sensitivity testing (NST) procedures. It is an easy procedure for a person to evaluate his/her progress.

Study of the acupuncture meridians is helpful to understand NST and how it works. If one learns to identify abnormal symptoms connected with acupuncture meridians, detection of the causative agents (allergens) will be easier. The pathological functions of the twelve major acupuncture meridians are given below.

THE LUNG MERIDIAN (LU)

Energy disturbance in the lung meridian affecting physical and physiological levels can give rise to the following symptoms:

Afternoon fever
Asthma between 3-5 a.m.
Atopic dermatitis
Bronchiectasis
Bronchitis
Burning in the eyes
Burning in the nostrils
Cardiac asthma
Chest congestion
Cough
Coughing up blood
Cradle cap
Dry mouth
Dry skin
Dry throat
Emaciated look
Emphysema
Fever with chills
Frequent flu-like symptoms
General body ache with
    burning sensation
Generalized hives
Hair loss
Hair thinning
Hay-fever
Headache between eyes
Inability to sleep after 3 a.m.
Infantile eczema
Infection in the respiratory
tract
Itching of the body
Itching of the nostrils
Itching of the scalp
Lack of desire to talk
Lack of perspiration
Laryngitis
Low voice
Moles
Morning fatigue
Mucus in the throat
Nasal congestion
Night sweats
Nose bleed
Pain between third   and
fourth thoracic vertebrae
Pain in the chest and
    intercostal muscles
Pain in the eyes
Pain in the first
    interphalangeal joint
Pain in the thumb
Pain in the upper arms
Pain in the upper back
Pain in the upper first and
    second cuspids (tooth)
Pharyngitis
Pleurisy
Pneumonia

Chapter 5

Poor growth of nails and hair
Postnasal drip
Profuse perspiration
Red cheeks
Red eyes
Restlessness between 3 to 5 a.m.
Runny nose with clear discharge
Scaly and rough skin
Sinus headaches
Sinus infections
Skin rashes
Skin tags
Sneezing
Sore throat
Stuffy nose
Swollen cervical glands
Swollen throat
Tenosynovitis
Thick yellow discharge in case of bacterial infection
Thin or thick white discharge in case of viral infection
Throat irritation
Tonsillitis
Warts

Energy disturbance in the lung meridian affecting the cellular level. When one fails to cry, when one feels deep sorrow, sadness will settle in the lungs and eventually cause various lung disorders.

Apologizing
Comparing self with others
Contempt
Dejection
Depression
Despair
False pride
Grief or sadness.
Highly sensitive emotionally
Hopelessness
Intolerance
Likes to humiliate others
Loneliness
Low self-esteem
Meanness
Melancholy
Over demanding
Over sympathy
Prejudice
Seeking approval from others
Self pity
Weeping Frequently

Essential Nutrients To Strengthen The Lung Meridian:

Clear water
Proteins
Vitamin A
Vitamin C
Bioflavonoid
Cinnamon
Essential fatty acids

Onions
Garlic
B-vitamins (especially B_2)
Citrus fruits
Green peppers
Black peppers
Rice

THE LARGE INTESTINE MERIDIAN (LI)

Energy disturbance in the large intestine meridian affecting physical and physiological levels can give rise to the following symptoms:

Abdominal pain

Acne on the face, sides of
   the mouth and nose

Asthma after 5 a.m.

Arthritis of the shoulder
   joint

Arthritis of the knee joint

Arthritis of the index
   finger

Arthritis of the wrist joint

Arthritis of the lateral part
   of the elbow and hip

Bad breath

Blisters in the lower gum

Bursitis

Dermatitis

Dry mouth and thirst

Eczema

Fatigue

Feeling better after a
   bowel movement

Feeling tired after a bowel
   movement

Flatulence

Inflammation of lower
   gum

Intestinal colic

Itching of the body

Loose stools or constipation

Lower backache

Headaches

Muscle spasms and pain of
lateral aspect of thigh, knee
   and below knee.

Motor impairment of the
fingers

 Pain in the knee

Pain in the shoulder,
   shoulder blade and back of
      the neck

Pain and swelling of the
   index finger

Pain in the heel

Sciatic pain

Swollen cervical glands

Skin rashes

Skin tags

Sinusitis

Tenosynovitis

Tennis elbow

Toothache

Warts on the skin.

Energy disturbance in the large intestine meridian affecting the cellular level can cause the following:

Guilt

Confusion

Brain fog

Bad dreams

 dwelling on past memory

Crying spells

Defensiveness

Inability to recall dreams

 Nightmares

 Nostalgia

Rolling restlessly in sleep

Seeking sympathy

Talking in the sleep and

Weeping

Essential Nutrients To Strengthen The Large Intestine Meridian:

Vitamins A, D, E, C, B, especially B_1, wheat, wheat bran, oat bran, yogurt, and roughage.

THE STOMACH MERIDIAN (ST)

Energy disturbance in the stomach meridian affecting physical and physiological levels can give rise to the following symptoms:

Abdominal Pains & distention
Acid reflux disorders
Acne on the face and neck
 ADD & ADHD
Anorexia
Autism
Bad breath
Black/ blue marks on the
  leg below the knee
Bipolar disorders
Blemishes
Bulimia
Chest muscle pain
Coated tongue
Coldness in the lower limbs
Cold sores in the mouth
Delirium
Depression
Dry nostrils
Dyslexia
Excessive hunger
Facial paralysis
Fever blisters
Fibromyalgia
Flushed face
Frontal headache
Herpes

Heat boils (painful acne) in
the upper front of the body
Hiatal hernia
High fever
Learning disability
Insomnia due to nervousness
Itching on the skin & rashes
Migraine headaches
Manic depressive disorders
Nasal polyps
Nausea
Nosebleed
Pain on the upper jaws
Pain in the mid-back
Pain in the eye
Seizures
Sensitivity to cold
Sore throat
Sores on the gums & tongue
Sweating
Swelling on the neck
Temporomandibular joint
   problem
Unable to relax the mind
Upper gum diseases
 Vomiting

Energy disturbance in the stomach meridian affecting the cellular level can cause the following:

Disgust
Bitterness
Aggressive behaviors
Attention deficit disorders
Butterfly sensation in the
  stomach
Constant thinking
Depression
Deprivation
Despair
Disappointment
Egotism

Emptiness
Greed
Hyperactivity
Manic disorder
Mental fog
Mental confusion
Nervousness
Nostalgia
Obsession
Paranoia
Poor concentration

Essential Nutrients To Strengthen The Stomach Meridian:

B complex vitamins especially B_{12}, B_6, B_3 and folic acid.

THE SPLEEN MERIDIAN (SP)

Energy disturbance in the spleen meridian affecting physical and physiological levels can give rise to the following symptoms:

Abnormal smell
Abnormal taste
Abnormal uterine bleeding
Absence of menstruation
Alzheimer's disease
Autism
Bitter taste in the mouth
Bleeding from the mucous
  membrane
Bleeding under the skin
Bruises under the skin
Carpal tunnel syndrome
Chronic gastroenteritis
Cold sores on the lips
Coldness of the legs
Cramps after the first day of
  menses
Depression
Diabetes
Dizzy spells
Dreams that make you tired
Emaciated muscles
Failing memory
Fatigue in general
Fatigue of the mind
Fatigued limbs
Feverishness
Fibromyalgia

Fingers and hands-numbness
Fluttering of the eyelids
Frequent
Generalized edema
Hard lumps in the abdomen
Hemophilia
Hemorrhoids
Hyperglycemia
Hypertension
Hypoglycemia
Inability to make decisions
Incontinence of urine or stool
Indigestion
Infertility
Insomnia: usually unable to fall
asleep
Intractable pain anywhere in
  the body
Intuitive and prophetic
  behaviors
Irregular periods
Lack of enthusiasm
Lack of interest in anything
Lethargy
Light-headedness
Loose stools
Nausea
Obesity

Pain and stiffness of the fingers
Pain in the great toes
Pallor
Pedal edema
Pencil-like thin stools with undigested food particles
Poor memory
Prolapse of the bladder
Prolapse of the uterus
Purpura
Reduced appetite
Sand-like feeling in the eyes
Scanty menstrual flow
Sensation of heaviness in the body and head
Sleep during the day
Slowing of the mind
Sluggishness
Schizophrenia
Stiffness of the tongue
Sugar craving
Swelling anywhere in the body
Swellings or pain with swelling of the toes and feet
Swollen eyelids
Swollen lips
Tingling or abnormal sensation in the tip of the fingers and palms
Varicose veins
Vomiting
Watery eyes

Energy disturbance in the spleen meridian affecting the cellular level can cause the following:

Anxiety
Concern
Does not like crowds
Easily hurt
Gives more importance to self
Hopelessness
Irritable
Keeps feelings inside
Lack of confidence
Likes loneliness
Likes to be praised
Likes to take revenge
Lives through others
Low self esteem
Needs constant encoragement
Obsessive compulsive behavior
Over sympathetic to others
Unable to make decisions
Restrained
Shy/timid
Talks to self
Worry

Essential Nutrients To Strengthen The Spleen Meridian:

Vitamin A, vitamin C, calcium, chromium, protein, berries, asparagas, bioflavonoids, rutin, hesparin, hawthorn berries, oranges, root vegetables, and sugar.

THE HEART MERIDIAN (HT)

Energy disturbance in the heart meridian affecting the Physical and physiological level can cause the following:

Angina-like pains
Chest pains
Discomfort when reclining
Dizziness
Dry throat
Excessive perspiration
Feverishness
Headache
Heart palpitation
Insomnia—unable to fall asleep
When awakened in the middle
  of sleep

Heaviness in the chest
Hot palms and soles
Irritability
Mental disorders
Nervousness
Pain along the left arm
Pain along the scapula
Pain and fullness in the chest
Pain in the eye
Poor circulation
Shortness of breath
Shoulder pains

Energy disturbance in the heart meridian affecting the cellular level can cause the following:

Joy
Lack of joy

Self-confidence
Compassion and love

Abusive nature
Aggression
Anger
Bad manners
Compulsive behaviors
Does not like to make
friends
Does not trust anyone
Easily upset

Excessive laughing or crying
Guilt
Hostility
Insecurity
Lack of emotions/
overexcitement
Lack of love and compassion
Sadness
Type A personality

Essential Nutrients To Strengthen The Heart Meridian:

Calcium, vitamin C, vitamin E, fatty acids, selenium, potassium, sodium, iron, and B complex.

THE SMALL INTESTINE MERIDIAN (SI)

Energy disturbance in the small intestine meridian affecting physical and physiological levels can give rise to the following symptoms:

Abdominal fullness
Abdominal pain
Acne on the upper back
Bad breath
Bitter taste in the mouth
Constipation
Diarrhea
Distention of lower abdomen
Dry stool
Frozen shoulde
Knee pain
Night sweats

Numbness of the back of the
shoulder and arm
Numbness of the mouth and
tongue
Pain along the lateral aspect
of the shoulder and arm
Pain in the neck
Pain radiating around the waist
Shoulder pain
Sore throat
Stiff neck

Energy disturbance in the small intestine meridian affecting the cellular level can cause the following:

Insecurity
Absentmindedness
Becoming too involved with
details
Day dreaming
Easily annoyed
Emotional instability
Feeling of abandonment
Feeling shy
Having a tendency to be
introverted and easily hurt

Irritability
Excessive joy or lack of
joy
Lacking- confidence
Over excitement
Paranoia
Poor concentration
Sadness
Sighing
Sorrow
Suppressing deep
sorrow

Essential Nutrients To Strengthen The Small Intestine Meridian:

Vitamin B complex, vitamin D, vitamin E, acidophilus, yoghurt, fibers, fatty acids, wheat germ and whole grains.

THE URINARY BLADDER MERIDIAN (UB)

Energy disturbance in the bladder meridian affecting physical and physiological levels can give rise to the following symptoms:

Arthritis of the joints of little finger
Bloody urine
Burning urination
Chills
Chronic headaches at the back of the neck
Disease of the eye
Enuresis
Fever
Frequent urination
Headaches especially at the back of the neck
Loss of bladder control
Mental disorders
Muscle wasting
Nasal congestion
Pain and discomfort in the lower abdomen
Pain in the inner canthus of the eyes
Pain behind the knees
Pain and stiffness of the back
Pain in the fingers and toes
Pain in the lateral part of the sole
Pain in the lower back
Pain along back of the leg and foot
Pain in the lateral part of the ankle
Pain along the meridian
Pain in the little toe
Painful urination
Retention of urine
Sciatic neuralgia
Spasms along the posterior part of the thigh and leg
Spasm behind the knee
Spasms of the calf muscles
Stiff neck
Weakness in the rectum and rectal muscle

Energy disturbance in the bladder meridian affecting the cellular level can cause the following:

Fright
Sadness
Disturbing and impure
  thoughts
Annoyed
Fearful
Unhappy

Frustrated
Highly irritable
Impatient
Inefficient
Insecure
Reluctant
Restless

Essential Nutrients To Strengthen The Bladder Meridian:

Vitamin C, A, E, B complex, especially B_1, calcium, amino acids and trace minerals.

THE KIDNEY MERIDIAN (KI)

Energy disturbance in the kidney meridian affecting physical and physiological levels can give rise to the following symptoms:

Bags under the eyes
Blurred vision
Burning or painful urination
Chronic diarrhea
Coldness in the back
Cold feet
Crave salt
Dark circles under the eyes
Dryness of the mouth
Excessive sleeping
Excessive salivation
Excessive thirst
Facial edema
Fatigue
Fever with chills
Frequent urination
Impotence
Irritability
Light- headedness
Lower backache

Motor impairment
Muscular atrophy of the
 foot
Nagging mild asthma
Nausea
Pain in the sole of the foot
Pain in the posterior aspect
 of the leg or thigh
Pain in the ears
Poor memory
Poor concentration
Poor appetite
Puffy eyes
Ringing in the ears
Sore throat
Spasms of the ankle and
 feet
Swelling in the legs
Swollen ankles and vertigo

Energy disturbance in the kidney meridian affecting the cellular level can cause the following:

Fear
Terror
Caution
Confused

Indecision
Paranoia
Seeks attention
Unable to express feelings

Essential Nutrients To Strengthen The Kidney Meridian:

Vitamins A, E, B, essential fatty acids, amino acids, sodium chloride (table salt), trace minerals, calcium and iron.

THE PERICARDIUM MERIDIAN (PC)

Energy disturbance in the pericardium meridian affecting physical and physiological levels can give rise to the following symptoms:

Chest pain
Contracture of the arm or elbow
Excessive appetite
Fainting spells
Flushed face
Frozen shoulder
Fullness in the chest
Heaviness in the chest
Hot palms and soles
Impaired speech
Irritability
Motor impairment of the tongue
Nausea
Nervousness
Pain in the anterior part of the thigh
Pain in the eyes
Pain in the medial part of the knee
Palpitation
Restricting movements
Sensation of hot or cold
Slurred speech
Spasms of the elbow and arm

Energy Disturbance In The Pericardium Meridian Affecting The Cellular Level Can Cause The Following:

Extreme joy
Fear of heights
Heaviness in the chest due
  to emotional overload
Heaviness in the head
Hurt
Imbalance in sexual energy
  like never having enough
  sex
In some cases no desire for
sex

Jealousy
Light sleep with dreams
Manic disorders
Over- excitement
Regret
Sexual tension
Shock
Stubbornness
Various phobias

Essential Nutrients To Strengthen The Pericardium Meridian:

Vitamin E, Vitamin C, Chromium, Manganese, Lotus seed, and Trace Minerals.

THE TRIPLE WARMER MERIDIAN (TW)

Energy disturbance in the triple warmer meridian affecting physical and physiological levels can give rise to the following symptoms:

Abdominal pain
Always feels hungry even after eating a full meal
Constipation
Deafness
Distention
Dysuria
Edema
Enuresis
Excessive thirst
Excessive hunger
Fever in the late evening
Frequent urination
Indigestion
Hardness and fullness in the lower abdomen
Pain in the medial part of the knee
Pain in the shoulder and upper arm
Pain behind the ear
Pain in the cheek and jaw
Redness in the eye
Shoulder pain
Swelling and pain in the throat
Vertigo

Energy disturbance in the triple warmer meridian affecting the cellular level can cause the following:

Depression
Deprivation
Despair
Emptiness
Excessive Emotion
Grief
Hopelessness
Phobias

Essential nutrients to strengthen the triple warmer meridian:

Iodine, trace minerals, vitamin C, calcium, fluoride, radish, onion, zinc, vanadium, and water.

THE GALL BLADDER MERIDIAN (GB)

Energy disturbance in the gall bladder meridian affecting physical and physiological levels can give rise to the following symptoms:

A heavy sensation in the right upper part of the abdomen
Abdominal bloating
Alternating fever and chills
Ashen complexion
Bitter taste in the mouth
Burping after meals
Chills
Deafness
Dizziness
Fever
Headaches on the sides of the head
Heartburn after fatty foods
Hyperacidity
Moving arthritis
Pain in the jaw

Nausea with fried foods
Pain in the eye
Pain in the hip
Pain and cramps along the anterolateral wall
Poor digestion of fats
Sciatic neuralgia
Sighing
Stroke-like condition
Swelling in the submaxillary region
Tremors
Twitching
Vision disturbances
Vomiting
Yellowish complexion

Energy Disturbance In The Gall Bladder Meridian Affecting The Cellular Level Can Cause The Following:

Aggression
Complaining all the time
Rage

Fearful, finding faults with others
Unhappiness.

Essential Nutrients To Strengthen The Gall Bladder Meridian:

Vitamin A, apples, lemon, calcium, linoleic acids and oleic acids (for example, pine nuts, olive oil).

THE LIVER MERIDIAN (LIV)

Energy disturbance in the liver meridian affecting physical and physiological levels can give rise to the following symptoms:

Abdominal pain
Blurred vision
Dark urine
Dizziness
Enuresis
Bright colored bleeding during menses
Feeling of obstruction in the throat
Fever
Hard lumps in the upper abdomen
Headache at the top of the head
Hernia
Hemiplegia
Irregular menses
Jaundice
Loose stools
Pain in the intercostal region
Pain in the breasts
Pain in the lower abdomen
Paraplegia
PMS
Reproductive organ disturbances
Retention of urine
Seizures
Spasms in the extremities
Stroke-like condition
Tinnitus
Vertigo
Vomiting.

Energy disturbance in the liver meridian affecting the cellular level can cause the following:

Anger
Irritability
Aggression
Assertion
Rage
Shouting
Talking loud
Type A personality

Essential Nutrients To Strengthen The Liver Meridian:

Beets, green vegetables, vitamin A, trace minerals, vitamin F

THE GOVERNING VESSEL MERIDIAN (GV)

Energy disturbance in the governing vessel meridian affecting physical, physiological and psychological levels can give rise to various mixed symptoms of other yang meridians.

This channel supplies the brain and spinal region and intersects the liver channel at the vertex. Obstruction of its Chi may result in symptoms such as stiffness and pain along the spinal column. Deficient Chi in the channel may produce a heavy sensation in the head, vertigo and shaking. Energy blockages in this meridian (which passes through the brain) may be responsible for certain mental disorders. Febrile diseases are commonly associated with the governing vessel channel and because one branch of the channel ascends through the abdomen, when the channel is unbalanced, its Chi rushes upward toward the heart. Symptoms such as colic, constipation, enuresis, hemorrhoids and functional infertility may result.

THE CONCEPTION VESSEL MERIDIAN (CV, REN)

Energy disturbance in the conception vessel meridian affecting physical, physiological and psychological levels can give rise to various mixed symptoms of other yin meridians.

The conception vessel channel is the confluence of the Yin channels. Therefore, abnormality along the conception vessel channel will appear principally in pathological symptoms of the Yin channels, especially symptoms associated with the liver and kidneys. Its function is closely related with pregnancy and, therefore, has intimate links with the kidneys and uterus. If its Chi is deficient, infertility or other disorders of the urogenital system may result. Leukorrhea, irregular menstruation, colic, low libido, impotency, male and female infertility are associated with the conception vessel channel.

Any allergen can cause blockage in one or more meridians at the same time. If it is causing blockages in only one meridian, the patient may demonstrate symptoms related to that particular meridian. The intensity of the symptoms will depend on the severity of the blockage. The patient may suffer from one symptom, many symptoms or all the symptoms of this blocked meridian. Sometimes a patient can have many meridians blocked at the same time. In such cases, the patient may demonstrate a variety of symptoms, one symptom from each meridian or many symptoms from certain meridians and one or two from other. Some patients with blockage in one meridian can demonstrate just one symptom from the list, but it may be with great intensity.

Some people, even though they have energy disturbances in multiple meridians, may not show any symptoms. Such patients might have a better immune system than others. Variations with all these possibilities can make diagnosis difficult in some cases.

Freedom From Environmental Sensitivities

6

NAET Testing Procedures

When an allergen's incompatible energy comes close to a person's energy field, repulsion takes place. Without recognizing this repulsive action, we frequently go near allergens (whether they are foods, drinks, chemicals, environmental substances, animals or humans) and interact with their energies. This causes energy disturbance in the meridians. These disturbances cause imbalances in the body, which cause illnesses, that create disorganization in body functions. This disorganization of the body and its functions involves the vital organs, their associated muscle groups and nerve roots, and possibly gives rise to chemical or environmental sensitivities and other allergy-related disorders.

To prevent the allergen from causing further disarray after producing the initial blockage, the brain sends messages to every cell of the body to reject the presence of the allergen. This rejection will appear as repulsion, the repulsion between energies will create weakness in the person's body and this weakness will produce different unpleasant health-related symptoms like pains, fatigue, irritability, general weakness, breathing difficulty, etc. The manifesting symptoms and their intensities will depend on the type of body tissue or organs involved, the amount of area involved, and the status of the person's immune system.

Your body has an amazing way of telling you when you are in trouble. As a matter of habit, though, you often have to be hurting severely before you look for help. If you went for help at the earliest hint of need, you would save yourself from unnecessary pain and agony. This applies to environmental allergies, too. If you identify your environmental allergens before you are exposed to them, you won't have to suffer the consequences. If you understand your body, your brain and their clues, you can avoid the causes that contribute to energy disturbances and body imbalances.

When people go near environmental allergens, they receive various clues from the brain, such as: an itchy throat, watery eyes, sneezing attacks, coughing spells, unexplained pain anywhere in the body, yawning, sudden tiredness, etc. You can demonstrate these changes by testing the strength of any part of the body in the presence and absence of the environmental allergens. A strong muscle of the arm, hand or leg can be used for this test. Test a strong muscle for its strength away from the allergen, then test it again in the presence of the allergen and compare the strength. The muscle will stay strong without any allergen near the body, but will weaken in the presence of an allergen. This response can be used to your advantage to demonstrate the presence of an allergen near you. We have done various studies on this subject and are in the process of publishing the studies.

NEUROMUSCULAR SENSITIVITY TESTING (NST)

Neuromuscular sensitivity testing (NST) is one of the tools used by NAET specialists to test imbalances in the muscles and the nervous system and in their communication pathways. The same neuromuscular sensitivity testing can also be used to detect the affect of allergens in the body by measuring the changes the allergens produce in the muscle and nerve function. Because it is measuring the change in the muscles strength in association with the integrity of the nervous system in the presence and absence of allergens, the name neuromuscular sensitivity testing is more appropriate for this test.

NST can be performed in the following ways (Illustrations of different types of neuromuscular sensitivity testing can be seen on the following pages).

1. Standard NST can be done in standing, sitting or lying positions. You need two people to do this test: the person who is testing, the "tester," and the person being tested, the "subject."(see below.)

2. The "Oval Ring Test" can be used in testing yourself, and on a very strong person with a strong arm.

3. Surrogate testing can be used to test an infant, invalid person, extremely strong or weak person, or an animal. The surrogate's muscle is tested by the tester, subject maintains skin-to-skin contact with the surrogate while being tested. The surrogate does not get affected by the testing. NAET treatments can also be administered through the surrogate very effectively without causing any interference with the surrogate's energy.

STANDARD NST
(Neuromuscular Sensitivity Testing)

Two people are required to perform standard NST: the tester, and the subject. The subject can be tested lying down, standing or sitting. The lying-down position is the most convenient for both tester and subject; it also achieves more accurate results.

Step 1:   The subject lies on a firm surface with the left arm raised 45-90 degrees to the body with the palm facing outward and the thumb facing toward the big toe.

**FIGURE 6-1
STANDARD NEUROMUSCULAR
SENSITIVITY TESTING (NST)**

FIGURE 6-2
NST WITH ALLERGEN

Step 2: The tester stands on the subject's (right) side. The subject's right arm is kept to his/her side with the palm either kept open to the air, or in a loose fist. The fingers should not touch any material, fabric or any part of the table the arm is resting on. This can give wrong test results. The tester's left palm is contacting the subject's left wrist (Figure 6-1).

Step 3: The tester using the left arm tries to push the raised arm toward the subject's left big toe. The subject resists the push. The arm (indicator muscle) remains strong if the subject is well balanced at the time of testing. It is essential to test a strong predetermined muscle (PDM) to get accurate results. If the muscle of the raised arm is weak and gives way under pressure without the presence of an allergen, the subject is not balanced, or the tester might be trying to overpower the subject. The subject does not need to gather up strength from other muscles in the body to resist the tester. Only 5 to 10 pounds of pressure need to be applied for three to five seconds.

If the muscle shows weakness, the tester will be able to judge the difference with only that small amount of pressure.

Step 4: This step is used if the patient is found to be out of balance as indicated by the PDM (predetermined or indicator muscle) presenting weak–without the presence of an allergen. The tester then uses the balancing points by placing the fingertips of right hand at Point 1. The left hand is placed on Point 2 (see below and fingure 6-3). The tester massages these two points gently clockwise with the fingertips about 20-30 seconds, then repeats steps 2 and 3. If the PDM tests strong, continue on to step 5.

POINT 1: SEA OF ENERGY (See figure 6-3)

Location: Two fingerbreadths below the navel, on the midline. According to Oriental medical theory, this is where the energy of the body is stored in abundance. When the body senses any

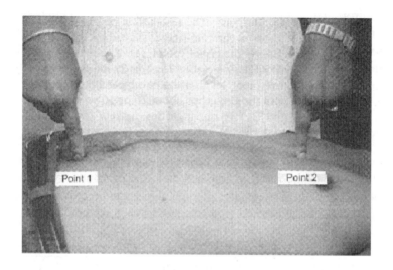

FIGURE 6-3
BALANCING THE PATIENT

danger around its energy field or when the body experiences energy disturbances, the energy supply is cut short and stored here. If you massage clockwise on that energy reservoir point, the energy will come out of this storage and travel to the part of the body where it is needed.

POINT 2: ENERGY CONTROLLER (see figure 6-3)

Location: In the center of the chest on the midline of the body, level with the fourth intercostal space. This is the energy dispenser unit. When the energy rises from the *Sea of Energy*, it goes straight to the *Energy Controller* point. This is the point that controls and regulates the energy circulation or Chi, in the body. From here, the energy is dispersed to different meridians, organs, tissues and cells as needed to help remove the energy disturbances. It does this by forcing energy circulation from inside out. Continue this procedure for 30 seconds to one minute and retest the NST. If the NST is found weak repeat the procedure until it gets strong. Check NST every 30 seconds.

Step 5: If the PDM remains strong when tested–a sign that the subject is balanced - then the tester should put the suspected allergen into the palm of the subject's resting hand. When the subject's fingertips touch the allergen, the sensory receptors sense the allergen's charges and relay the message to the brain. If it is an incompatible charge, the strong PDM will go weak. If the charges are compatible to the body, the indicator muscle will remain strong. This way, you can test any number of items to determine the compatible and incompatible charges.

Much practice is needed to test and sense the differences properly. If you can't test properly or effectively the first few times, don't get discuraged. Pactice makes perfect.

"OVAL RING TEST" OR "O RING TEST"

The "Oval Ring Test" or "O Ring Test" can be used in self-testing. This can also be used to test a subject, if the subject is physically very strong with a strong arm and the tester is a physically weak person. (See figure 6-4)

Step 1: The tester makes an "O" shape by opposing the little finger and thumb on the same hand. Then, with the index finger of the other hand he/she tries to separate the "O" ring against pressure. If the ring separates easily, the tester needs to be balanced as described above.

FIGURE 6-4
`O' RING TEST TO DETECT ALLERGIES

Step 2: If the "O" ring remains inseparable and strong, hold an allergen in the other hand, by the fingertips, and perform step 1 again. If the "O" ring separates easily, the person is allergic to the substance in the hand. If the "O" ring remains strong, the substance is not an allergen.

The finger-on-finger test (Figure 6-5) is another way to test yourself. The strength of the interphalangeal muscles of two fin-

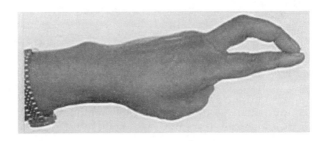

FIGURE 6-5
FINGER ON FINGER TEST

gers of one hand is used here to test and compare the strength without and with holding an allergen. The middle finger is pushed down, using the index finger of the same hand, or vice versa, in the absence and presence of the allergen in the other hand. This also needs much practice to become good at testing.

Step 1: The tester places the pad of the index finger at the back of the middle finger of the same hand. The middle finger is pushed down, using the index finger of the same hand. If the middle finger could resist the push by the index finger, then the person is balanced. If the person is not balanced, please balance using the same step-5 from standard NST. If the person is balanced, go to the next step.

Step-2: Then the tester holds the allergen in one hand. Then he/she places the pad of the index finger at the back of the middle finger of the same hand again. The middle finger is pushed down, using the index finger of the same hand. The item you are holing is an allergen if the middle finger goes down easily while pushing with the index finger in the presence of the item.

If done properly, NST is one of the most reliable methods of allergy tests, and it is fairly easy to learn and practice in every day life. It cuts out expensive laboratory work.

After considerable practice, some people are able to test very efficiently using these methods. It is very important for allergic people to learn some form of self-testing technique to screen out possible allergens from their daily life to prevent repeated allergic reactions on a daily basis. After receiving the basic 30-40 treatments from a NAET practitioner, a person can test and avoid unexpected allergens. Hundreds of new allergens are thrown into the world daily by people who do not understand the predicament of allergic people. If you or your child want to live in this world looking and feeling normal among people, side by side with the allergens, you need to learn how to test on your own. It is not practical for people to treat thousands of allergens from their surroundings or go to an NAET practitioner every day for life. If you learn to detect your allergies on your own after treating for NAET Basics, you can live with far fewer health problems.

A TIP TO MASTER SELF-TESTING

Find two items, one that you are allergic to and another that you are not, for example an apple and a banana.

You are allergic to the apple and not allergic to the banana. Hold the apple in the right hand and do the "Oval Ring Test" as shown in the figure 6-4. The ring easily breaks. The first few times if it didn't break, make it happen intentionally. Now hold the banana and do the same test. This time the ring doesn't break. Put the banana down, rub your hands together for 30 seconds. Take the apple and repeat the testing. Practice this until you can sense the difference. When you can feel the difference between these two items you can test anything around you.

SELF-TESTING PROCEDURE

Step-1: Find two items, or two group of items (collect a few samples of allergens to which you know you are allergic, perhaps items tested for you by another person). Then collect another group that you are not allergic to. Let's assume that you are allergic to the apple group and not allergic to the banana group in the following example.

Step-2: Hold the samples from apple group one at a time in one hand and test with the other hand, using either "O" ring testing or finger on finger testing. The ring easily breaks in the case of "O" ring testing or the interphalangeal muscle weakens easily if you are using finger-on-finger testing. The same way, test each item from the banana group one by one. The "O" ring remains strong and the middle finger will not be pushed down by the index finger when you test from the items from the non-allergic group.

Rub your hands together or wash your hands between touching different samples for testing.

Step-3: When you test the allergic items, if the muscle doesn't go weak, make it happen intentionally for the first few times. Now, hold the items from the non-allergic group and do the same test. This time, the ring doesn't break. Practice this procedure for awhile. Rub your hands together for 30 seconds between changing the test samples to interrupt the energy at the fingertips of the previous sample. Practice this as long as necessary until your subconscious mind is able to recognize the strength of the allergen just by touching it with your fingertips. When you master this procedure, you can test anything around you.

SURROGATE TESTING

This method can be very useful to test and determine the allergies of an infant, a child, a hyperactive child, an autistic child, disabled person, an unconscious person, an extremely strong, or a very weak person. You can also use this method to test an animal, plant, or a tree.

Three people are needed for this test, as shown in Figure 6-6. NAET treatments can be administered through the surrogate very effectively without causing any interference to the surrogate's energy.

The surrogate's muscle is tested by the tester. It is very important to remember to maintain skin-to-skin contact between the surrogate and the subject during the procedure. If you do not, then the surrogate will receive the results of testing and treatment.

The testing or treatment does not affect surrogate as long as the subject maintains uninterrupted skin-to-skin contact with the surrogate.

As mentioned earlier, muscle response testing is one of the tools used by kinesiologists. Practiced in this country since 1964, it was originated by Dr. George Goodheart and his associates. Dr. John F. Thie advocates this method through the "Touch For Health"

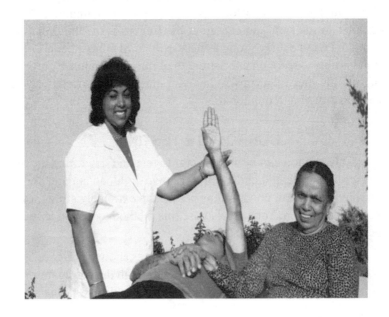

FIGURE 6- 6
TESTING THROUGH A SURROGATE

Foundation in Malibu, California. For more information and books available on the subject, interested readers can write to "Touch For Health" Foundation.

Muscle response testing can be used to test any substance for allergies. Even human beings can be tested for each other in this manner. When allergic to another human (father, mother, son, daughter, grandfather, grandmother, spouse, caretaker, baby sitter, etc.) you or your child could experience similar symptoms as you would with foods, chemicals or fabrics.

ABOUT PERSON-TO-PERSON ALLERGIES

If people are allergic to each other, the allergy can affect them in various ways: The father and/or mother allergic to each other, or to the child, or child allergic to a parent, the person who is allergic here can get sick or remain sick indefinitely. If the husband is allergic to the wife or wife towards the husband, they might fight all the time and/or their health can be affected. The same things can happen among other family members. It is important to test family members and other immediate associates for possible allergy and, if found, they should be treated for each other.

TESTING PERSON-TO-PERSON ALLERGIES

The subject lays down and touches the other person (Figure 6-7). The tester pushes the arm of the subject in steps 2 and 3 above. If the subject is allergic to the second person, the indicator muscle goes weak. If the subject is not allergic, the indicator muscle remains strong.

PETS'S ALLERGY

You can also check your allergy to your pets, and the pet's allergy towards you.

You can test your pet's allergy using this same method. You can treat the pets through a surrogate for their allergy to the sub-

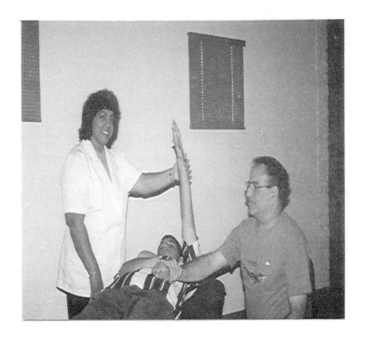

FIGURE 6- 7
TESTING SON'S ALLERGY TO THE FATHER

stances. Pets could be found allergic to their daily foods, drinks, tap water, toys, living environment, rug or mat where it sleeps, vitamins, drugs, vaccination, insects, tics, fleas, fungus, collar, cat liter, etc.

Many of my patients who own animals have been found suffer from fibromyalgia, vulvodynia, chronic fatigue, yeast infection, eczema, dermatitis, other skin problems, insomnia, repeated bladder infections, brain fog, briain irritability, digestive disorders, etc. They have been found allergic to animal food (dog food and cat food, which they expose to while opening the containers to feed them), insects, like fleas, tics, parasites, fungus, etc. Once they get treated for all these allergens through NAET, their symptoms get reduced or eliminated completely. I have detected these allergens via NST and tried testing them via standard laboratory procedures to detect allergens but was not very successful to duplicate the NST results with laboratory works. One of the explanation for this may be that energy imbalances take place in the body at the very first event of the energy interference and continue to get accumulated for a while before it can cause cell damage and tissue damage in the body. All laboratory works and imagimg works need certain amount of cell and tissue involvement before they can reproduce the damage in the blood tests, biopsies and X-rays. But when the energy interferences happen, when the body's enegy channels try to fight the invading energies from entering the body, body produces strong reactions which we call allergic reactions or later diseases. I think because of the above reason, such reactions from fighting the foreign energies (fleas, insects, animal epithelia, etc.) are difficult to reproduce in laboratory works. If people get treated for the animals and animal associated materials at the very first sign of the symptoms, probaly they will not become sick in the future.

TESTING OWNER 'S ALLERGY TO THE ANIMAL
Procedure:

Step-1: The owner's muscle is tested by the tester while the he/she maintains the skin-to-skin contact on the pet with the free hand. It is very important to remember to maintain skin-to-skin contact between the owner and the pet.

Step-2: The owner holds the pet (the allergen) with the fingers resting on the animals body while the tester pushing on the raised hand. If the NST is weak, the owner is allergic to the pet. The owner can get treated for the animal by maintaining the contact on its body, while the doctor is administering the treatment.

TREATING PET TOWARDS THE OWNER

Step-1: In this case the owner will be the surrogate. The owner's muscle is tested by the tester while the he/she maintains the skin-to-skin contact on the pet with the free hand. It is very important to remember to maintain skin-to-skin contact between the owner and the pet.

Step-2: The owner holds the pet (the allergen) with the animal's paw touching the body of the owner. The tester pushes on the raised arm of the owner while the animal's paw is resting on the owner's arm on the opposite side of the raised arm. Fingers resting on the animals body while. If the NST is weak, the owner is allergic to the pet. The owner can

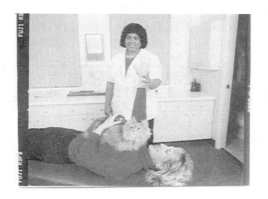

FIGURE 6-8
TESTING A PET'S ALLERGY TO FOOD

FIGURE 6-9
TESTING FOR AN ALLERGY TO THE PET

get treated for the animal by maintaining the contact on its body, while the doctor is administering the treatment.

NST should be taught in every medical school. It could be even taught to senior students and teachers in regular schools and other establishments. Everyone should learn to test to detect their allergies even if treatment is not available. If you know your allergies, you can avoid many unexpected health emergencies.

Medical professionals as well as the general public should be educated to look for various types of allergies. Parents should form support groups to encourage other parents to learn NST. Schools should begin teaching NST procedures to children as normal learning curriculum. Community centers, adult schools, hospitals involved in patient education sessions should teach NST to all their attendees. Allergies and allergy-related problems are on the rise due to today's scientific advancements. If we learn to manage our adverse reactions towards the allergens in our lives, we can enjoy the benefit of modern inventions and the wonderful world around us.

The knowledge of energy blockages and resultant diseases comes from Oriental medicine. Oriental medicine also teaches that, if given a chance and with a little support, the body will heal itself.

7

NAET Home-Help

T he purpose of this book is to inform people about NAET allergy elimination treatment, so that needy patients can learn about the availability of such a treatment and, if interested, can locate the appropriate medical practitioners with proper NAET training to help eliminate their allergies.

Information regarding a few important acupressure points, and NAET energy balancing points are described in the following pages. These points and techniques, when used properly according to the accompanying instructions, might help to reduce or control your presenting acute allergic symptoms. If used properly, these points can also be helpful in emergency situations.

A few self-help energy balancing applications are also discussed in this chapter, with illustrations. These balancing techniques are safe to use on people at any age and in any condition. These procedures can be safely used in balancing animals too. When one maintains energy in a balanced state, the body may not experience any illness or adverse reactions. Just by balancing the body regularly, by maintaining a balanced state, many people have reported that they were able to keep their allergic reactions under control. Some have reported reduction in their allergy-related other health conditions as well.

But again I would like to make the reader aware that these are only energy balancing techniques and should not be confused with actual NAET treatment procedures done with a trained NAET specialist. These balancing techniques will not replace the need for a trained practitioner. These techniques alone are not sufficient to permanently eliminate your allergies. These procedures, when used properly as described in the following page, will: help to improve overall health, reduce allergies and allergic reactions, help with allergy-related health problems, but will not eliminate your allergies.

TESTING

In Chapter 6, you learned to test and find your allergies using NST. You have learned to test and identify allergens in general. If you want to be healthy, you are urged to practice these testing techniques and make a habit of testing everything you suspect before exposing yourself to them. When you identify your allergens, you may be able to avoid them easily.

I spent countless hours testing, determining, researching, and trying out all my NAET discoveries on hundreds of people before I began sharing them with others. I was a desperate patient myself some time ago. I was told to learn to live with my chronic pain for the rest of my life. So I understand the pain of living with sickness and feeling trapped.

Now, we have a simple, safe, inexpensive, uncomplicated procedure to test allergies and allergy-related disorders with maximum accuracy (refer to JNECM, 2005).

You can test any type of allergen in this fashion. I am going to list a few commonly encountered, unsuspected, allergens below. When I tell someone to test each and every item before using, most people do not understand that one could be allergic to a vast number of everyday items around them. Most people, including some practitioners, miss unsuspected hidden allergens. Then people continue to suffer from various health problems and eventually, when their finances, spirits, motivation and hopes get exhausted, join the club of "Victims of Incurable Disorders."

TREATMENT

We treat many acute pains using NAET. But soon after the acute pain or symptoms are treated, we encourage our patients to return to the clinic to begin NAET Basic treatments. NAET Basic treatments are one's basic essentials for survival. Basic essentials are described in Chapter 4. When patients are treated for Basic essentials, they improve their immune system and can easily maintain a good immune system. Allergies and allergy-related diseases tend not to manifest in people with good immune systems.

We have found from our experience that patients who get treated just for acute symptoms or presenting symptoms like headaches, backaches, abdominal pains, etc., on the initial couple of visits and never return for NAET Basic essentials, continue to suffer various other sicknesses throughout their lives. Because of this we encourage you to complete at least the NAET Basic 15.

COMMONLY SEEN ALLERGENS AROUND YOU

After-shave lotion, razor-blades

Animals, their epithelial and dander

Bed, bed linen, bed sheet, comforter, and blanket

Books, papers, pens and pencils

Carpets and drapes

Ceramic cups and tiles on the floor

Freedom From Environmental Sensitivities

Child's school work materials

Clothing, bath towels and other fabrics, name tags

Colored clothes (people can be allergic to different colors not just the inks and dyes.)

Coloring books

Computer screen, keyboard, desk, and chair

Cooking dishes

Dishwashing soaps and scrubbers

Drinking water and tap water

Drinks

Eating utensils like plates, spoons, forks

Fruits and vegetables

Newspaper, prints, ink

Nightgowns, Pajamas

Grains and breads

Hair shampoo, hair conditioner and body lotions

Housecleaning products

Latex gloves and office products

Leaves, weeds, grass, and flowers

Oils, and other food items

Newspaper and ink

Pillow and pillow case

Toothbrush, toothpaste, mouthwash, dental floss

Toothpick, Q-tips, other hygienic materials

Toys, stuffed animals

Vitamins and drugs

Work materials

NAET SELF-TESTING PROCEDURE #1
HOLD, SIT AND TEST

This is the most simple allergy testing procedure. We teach this to our patients during patient-education classes. This is very simple and our young patients love it. Children are thrilled by this procedure. They test secretly for their food, cookies, drinks, clothes, school books, etc., before the parents get to test them with NST.

MATERIALS NEEDED:

1. A sample holder (thin glass jar, test tube, or a baby food jar with a lid can serve as a sample-holder).

2. Samples of the suspected allergens.

All perishable items, liquids, foods, should be placed inside the jar, then the lid should be closed tightly so that the smell will not bother the patient. If it is a piece of fabric, toy, etc., it can be held in the hand. Severe allergens like pesticides, perfume, chemicals, other toxic products should only be self-tested by adults, never by children, although it is better for anything poisonous like pesticides, insecticides, fumigants to be tested by trained NAET personnel. One must be very careful with these.

PROCEDURE:

Place a small portion of the suspected allergen in the sample holder and hold it in your palm, touching the jar with the fingertips of the same hand for 15 to 30 minutes. If you are allergic to the item in the jar, you will begin to feel uneasy when holding the allergen, giving rise to various unpleasant allergic symptoms, or exacerbation of prior allergic symptoms. The intensity of symptoms experienced is directly related to the severity of the allergy.

When one holds an allergen, one or more of the symptoms from the following list may be experienced:

Abdominal discomforts
Anger
Asthma
Backaches
Begins to get hot or cold on various parts of the body
Blurry eyes
Brain fog
Butterfly sensation in the stomach
Chest pains
Cough
Cravings
Crying spells
Deafness or ringing in the ear
Dry mouth, nose or throat
Fatigue
Flatulence
Frequency of urination
Headaches
Heaviness in the head
Heaviness in the chest
Heavy sensation in the body
Hives
Hyperactivity
Insomnia
Irregular heart beats (fast or slow)
Irritability
Itching in the nose, eyes, cheeks or ears
Knee or other joint pains
Light-headedness
Migraine headaches
Mucus in the throat
Nausea

Nervousness
Nose bleeds
Pin prick sensation
Pins and needles on the palms or soles
Poor attention span
Poor bowel control
Poor vision
Rashes
Redness on the cheeks or ears
Restlessness
Runny nose or blocked nostrils
Shortness of breath
Sinus troubles
Sneezing attacks
Sudden appearance of canker sores
Sudden eruption of acne or pimples on the face or body
Suddenly becomes silent
Suddenly becomes talkative
Unexplained pain anywhere in the body
Watering from the eyes
Weakness of any part of the limbs

Since the allergen is inside the sample-holder when such uncomfortable sensations are felt, the allergen can be put away immediately and the person can wash his/her hands to remove the energy of the allergen from the fingertips. This should stop the reaction immediately. In this way, you can determine allergens and the degree of allergy easily without putting yourself in danger.

More accurate tests can be found in Chapter 6. The above method is a quick-find, and not everyone responds in the same way.

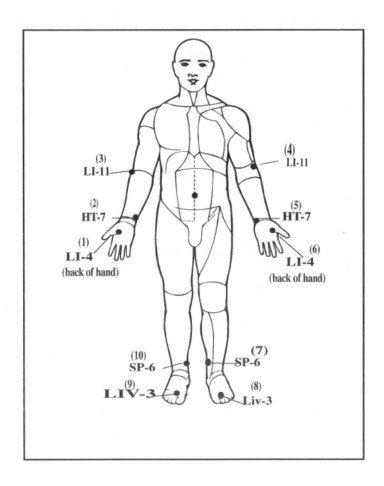

FIGURE 7-1
NAET BALANCING POINTS

NAET BALANCING TECHNIQUE

Look at the Figure 7-1. Massage the ten points gently for a minute each, clockwise, in circular motion with your fingerpads starting from point-1, go through pt-2, pt-3, pt-4, pt-5, pt-6, pt-7, pt-8, pt-9, pt-10, and finish up the massage again at point-1. A point stimulator may be used in place of finger massage. Massaging these points can improve the nerve energy circulation in the meridians. Treating these points will increase your overall energy. (After each NAET treatment in a practitioner's office, these points should be massaged or needled to balance the body in order to help the NAET treatment last.

If the person's reaction is mild, then he/she can hold an allergen while massaging the points. If the allergen is producing severe symptoms do not have the person hold the allergen while massaging. Just massaging these points will balance the person. For severe reactions, you must call emergency help.

Please do not treat severe reactions using this method. Find a NAET practitioner or go to the nearest emergency room.

ACUTE ALLERGIC REACTIONS

If you get an acute reaction to food you ate in a restaurant or at home, or in contact with an allergen that you can identify as the causative allergen, you may hold it in a sample-holder to balance your body using NAET balancing Points as in figure 7-1. Repeat the balancing treatments ten minutes apart until you feel better. By treating this way, you may not clear the allergy permanently but you could control your allergic symptoms temporarily. When stabilized, you should go to the appropriate medical practitioner for evaluation and treatment for lasting results.

THE ORIENTAL MEDICINE CPR POINT

The CPR (Cardio-Pulmonary Resucscitation) point of Oriental Medicine is the Governing Vessel-26. Please see figure7-2.

Location: Below the nose, a little above the midpoint of the philtrum. This is a very important area to know about in case of emergencies. Please memorize this information and teach your family and friends!

Indication: Fainting, sudden loss of consciousness, cardiac arrhythmia, heart attack, stroke, sudden loss of energy, hypoglycemia, heat-stroke, sudden pain in the lower back, general lower backache, breathing problem due to allergic reactions, mental confusion, mental irritability, anger, uncontrollable rage, exercise-induced anaphylaxis, anaphylactic reactions to allergens, and sudden breathing problem due to any cause.

Procedure: Massage or stimulate the point for 30 seconds to a minute at the beginning of the problem.

• If you are treating yourself to wake you up from feeling sleepy while driving, or to recover from sudden loss of energy, etc., massage gently on this point. For example: While you are driving, if you feel sudden loss of energy or sensation of fainting, immediately massage this point. Your energy will begin to circulate faster and you will prevent a fainting episode.

• If you are reviving a suddenly unconscious victim, massage the point vigorously to inflict slight pain so that the person will wake up immediately. Vigorous massage is used only to awaken a suddenly unconscious person or a person who became unresponsive in front of you. Vigorous stimulation may not produce the same result when a person has been unconscious for a long duration. This should not be used if injury to the face or upper gum is suspected.

POINTS TO HELP WITH MEDICAL EMERGENCIES

1. Fainting: GV-20, GV-26, GB-12, LI-1, PC-9, Kid-1

2. Nausea: CV-12, PC-6, Ht-7

3. Backache: GV-26, UB-40

4. Fatigue: CV-6, LI-1, CV-17

5. Fever: LI-11, GV-14

Stimulate these points by massaging them clockwise one at a time (from right to left on the patient's body) to control the acute symptoms as mentioned above. Patients will usually respond within 30 seconds to one minute of stimulating these points. If someone is slow to respond, it is OK to massage for three to five minutes. But please stop and evaluate the condition of the patient every 30 seconds. These methods are not meant as cures. If there is little or no response within a short time, the person might need medical help.

Do Not Hesitate To Call For Emergency Help (911).

For more information on revival points, refer to Chapter 3, pages 570 to 573, in "Acupuncture: A Comprehensive Text," by Shanghai College of Traditional Medicine, Eastland Press, 1981, or refer to my book, *Living Pain Free with Acupressure*, 1997, available at various bookstores and at our website **naet.com**

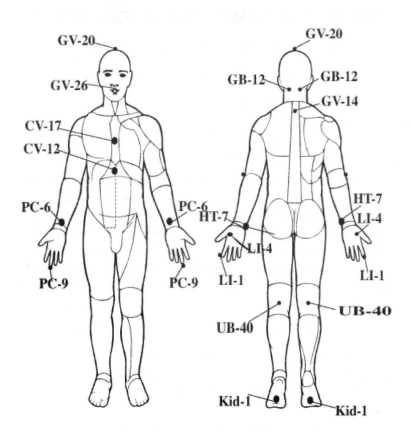

**FIGURE 7-2
RESUSCITATION POINTS**

EMOTIONAL IMBALANCES

People throughout the world suffer from various emotional blockages due to today's life-styles. People check each other in everything they do in their lives. No one is free from this check system.

Look at this simple example:

A mother sends her son to school. She pays a fee to the school to educate him. A teacher is assigned to teach the child. If the teacher did not teach the child, he would not know his test questions. If the child did not study, he would not pass the test. If he did not pass the test, he will not be promoted. If he continues to fail the test, his mother is going to take him out of the school. If the child did not attend the school, the school will not get paid. The teacher will not get paid. If the teacher did not get paid, he/she will not work there. The school will not function without money and a teacher.

Look at any aspect of the life-style around you. You can see a similar check system everywhere. Everyone is connected with one another by this check system. People who follow the check system closely win and live with less emotional blockages. People who do not follow the system carefully will swallow lots of emotions every day. These unresolved emotions will adhere to the gut wall and begin to eat it up, creating health disorders. Until you understand this and find a way to move emotion out of its established location (wherever it is rooted), you may not be healthy.

As you accumulate emotions, you need to eliminate them immediately. Do not collect them. Environmentally ill patients have very fragile emotional centers. They must eliminate their emotional blockages completely if they want to get healthy. Until now, many people talked about emotional issues, some people understood the importance of recognizing them; others thought these issues might lead to health disorders. But no one quite knew how to remove them. Now there is an easy, self-treatment method to clear your troubled emotions in the privacy of your home through NAET. This technique has been tried on many hundreds of patients in my office with excellent results. It is designed to help you free yourself from your entanglements and achieve your basic needs of life. Use it as freely and frequently as you need to and you will be able to attain your dreams!

METHOD

The best time to clear your day's emotions is each night before you go to sleep. However small an emotion may be, clear it on the same day it happens. This can prevent you from accumulating emotions and causing other health problems tagged on to the emotional issues in later life.

You may go to an NAET practitioner near you to remove your blockages with NAET, a slightly different procedure needing professional knowledge. You can get results with certain emotions using the following technique.

NAET EMOTIONAL SELF-TREATMENT

Find a calm and comfortable area where you can be alone with your thoughts without anyone disturbing you. You are advised to sit or lie down with your eyes closed. Closing your eyes will help you concentrate better. Do not do this procedure in a standing position. Some people may experience problems in maintaining a stable equilibrium in a standing position with the eyes closed.

METHODOLOGY:

Tap and clear (See Figure 7-3 for point locations).

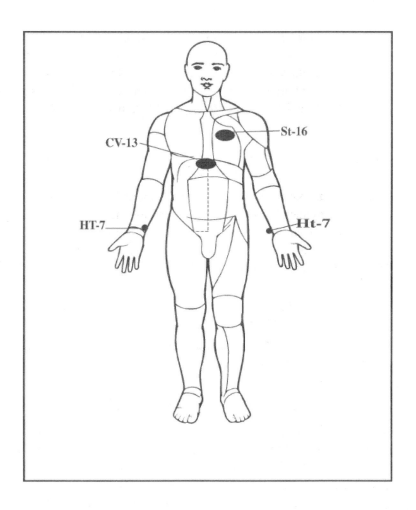

FIGURE 7-3
EMOTIONAL SELF-HELP POINTS

Step-1: Place your right three fingertips on the "pt-1" (pt-1 on the diagram) and left three fingertips on the "pt-2" (pt-2 on diagram 7-3).

While maintaining the left fingertips' contact on "pt-2," and tap on "pt-1," using right fingertips, while **reliving** your trauma, or incident of the day. The tapping time is 60 seconds on each point.

Step-2: Maintaining the right fingertips' contact on "pt-1," tap with the left fingertips on "pt-2," while **reliving** the memory of the incident.

Step-3: Massage clockwise on point "Rt Ht-7" then on "Lt. Ht-7" one minute each, while thinking positive or pleasant thoughts about the incident.

Do it every night before sleep as long as it takes to get you strong emotionally. If you do not have any particular incident to overcome, just massage these points without thinking anything particular. It is going to make your emotional health better anyway.

NUTRITIONAL SUPPORT
FOOD AND NERVOUS SYSTEM

Environmentally allergic people have sluggish liver, lung, spleen, large intestine function, which leads to poor energy circulation in those areas. Foods with warming properties (more alkaline food) should be eaten more often to improve the energy flow to the liver and spleen meridians and organs.

ALKALINE FOODS:

♦   Reduce inflammation.

♦   Promote less brain fog, can think and comprehend faster.

♦   Help to maintain normal circulation of blood, lymph and nerves.

♦   Help to maintain even distribution of energy throughout the body, Help reduce the allergic reactions.

♦   Helps to maintain overall calmness.

♦   People with alkaline bodies heal fast and tend to be less emotional.

ACIDIC FOODS:

♦   Cause more inflammation

♦   Cause more muscle aches, generalized pains and joint pains.

♦   Cause body and brain fatigue.

♦   People with acidic bodies are slow thinkers, and will have sluggish brain function.

♦   Cause Indigestion, poor assimilation and poor elimination.

♦   Cause poor circulation of blood, lymph and nerve energy.

♦   Wounds heal slowly in an acidic body.

♦   People with acidic bodies should eat more of alkaline foods and less of acidic foods.

♦   People with highly acidic bodies can be very emotional.

Alkaline and Acid food groups are given below. Try to eat more alkaline and less acidic foods. Please make sure you or your child are not allergic to these foods. If you find your child allergic to these foods, please see an NAET practitioner to get treated before eating.

Eggs are alkaline by nature.

Eggs Alkaline: Quail egg, duck egg.

Eggs Mildly Acidic: Chicken egg.

Human milk is alkaline.

Dairy Less Acidic: Aged Cheese, Butter, Cow Milk, Cream, Goat Cheese, Goat Milk, and Yogurt.

Dairy More Acidic: Cottage Cheese, Casein, Ice Cream, Milk protein, New Cheese, and Processed Cheese.

Vegetables - alkaline: Arugula, Asparagus, Beet, Bell Pepper, Broccoli, Brocoflower, Brussels Sprout, Burdock, Cabbage, Cauliflower, Celery, Chives, Cilantro, Collard Greens, Cucumber, Daikon, Egg Plant, Endives, Garlic, Ginger Root, Jicama, Kale, Kohlrabi, Lettuce, Lotus Root, Mung Bean, Mushroom, Mustard Greens, Onion, Parsley, Parsnip, Potato, Radish, Rutabaga,

Scalion, Seaweed and Other Sea Vegetables, Soybean, Squash, Sweet Potato, Taro Root, Turmeric, Turnip, and Yam.

Vegetables - Acidic: Carrot, Chard, Lima Bean, Navy Bean, Peanut, Rhubarb, Snow Pea, Spinach, and Zucchini.

Fruits - alkaline: Apple, Apricot, Avocado, Banana, Blackberry, Blueberry, Boysenberry, Cantaloupe, Cherry, Citrus, Grape, Grapefruit, Honeydew, Lemon, Lime, Loganberry, Mango, Nectarine, Olive, Orange, Papaya, Peach, Pear, Persimmon, Pineapple, Raspberry, Raw Tomato, Strawberry, Tangerine, and Watermelon.

Fruits - acidic: Canned Fruit, Cherimoya, Cooked Tomato, Cranberry, Date, Dry Fruit, Fig, Guava, Plum, Pomegranate and Prune.

Grains are acidic by nature but some are less acidic and some are more acidic. Try to eat less acidic grains.

Grains - Less Acidic: Amaranth, B vitamins, Brown Rice, Buck Wheat, Farina, Kamut, Kasha, Millet, Oat, Quinoa, Ragi, Sago, Semolina, Spelt, Wheat, White Rice, Wild Rice. Sago grains (produced from sago palm) is a natural tranquilizer. Try to consume them at night to induce good sleep.

Grains - More Acidic: All Purpose Flour, Barley, Barley Groat, Bleached Flour, Corn, Rye, Maize, And Oat Bran.

Meat is acidic by nature.

Meat - Less Acidic: Boar, Chicken Egg, Elk, Fish, Game, Gelatin, Goose, Lamb, Mollusks, Organ Meat, Shellfish, Turkey, Venison and wild duck.

Meat - More Acidic: Bear Meat, Beef, Chicken, Lobster, Mussel, Pheasant, Pork, Squid, and Veal.

Dried Bean - Alkaline: Lentils and mung beans

Dried Beans - Acidic: Azuki Beans, Black Eyed Peas, Fava Beans, Green Pea, Kidney Bean, Navy Bean, Pinto Bean, Chick Pea, Garbanzo Bean, Beans, Green Mung Beans, Lima Bean, Split Pea, Split Pea, String Bean, White Bean, Red Bean, and Soybean.

Fats - Alkalne: Avocado Oil, Borage Oil, Clarified Butter, Coconut Oil, Cod Liver Oil, Evening Primrose Oil, Flax Seed Oil, Hydrogenated Oil, Linseed Oil, and Olive Oil.

Fats - Less -Acidic: almond oil, canola oil, grapeseed oil, pumpkin seed oil, safflower oil, sesame oil, and sunflower oil.

Fats - More Acidic: Brazil Nut Oil, Chestnut Oil, Cottonseed Oil, Palm Oil, Soy Oil, and Superheated Vegetable Oils.

Nuts - Alkaline: Almond, Cashew, Chestnut, Poppy Seed, Pumpkin Seed, Sesame Seed.

Nuts - Acidic: Brazil Nut, Hazelnut, Peanut, Pecan, Pine Nut, Soy.

Drinks - alkaline: Apple cider vinegar, Cow Milk, Goat Milk, Herbal tea, Honey, Human milk, Maple syrup, Mineral water, Molasses, Rice milk, Rice Syrup, Sugar Cane juice, and Yogurt.

Drinks - acidic: Alcohol, Beer, Cane Sugar, Carbonated Drinks, Chocolate Drinks, Cocoa, Coffee, Cranberry Juice, Ice Cream, Malt, Rice Drink, Rice Milk, Rice Vinegar, Soy Milk, Table salt, White Sugar, White Vinegar, and Yeast,

Juices and drinks: Vegetable juices and broths are preferred to fruit juices. Fruit juices should be taken in minimum quantity (four to six ounces daily). Eating many fruits, and sugar products, drinking a lot of fruit juices etc., can encourage candida growth. After NAET, it is okay to have them in moderation. Herbal tea can be used as needed. Avoid carbonated drinks. Instead drink 5-6 glasses of purified or boiled cooled water daily. Make sure you are not allergic to it. (Yes, one can be allergic to certain waters).

Avoid artificial sweeteners, food colorings and additives as much as possible even after NAET treatments. Try to eat more organic, less chemically contaminated, less refined foods. Drink nonallergic water.

Freedom From Environmental Sensitivities

Glossary

Acetaldehyde: An aldehyde found in cigarette smoke, vehicle exhaust, and smog. It is a metabolic product of Candida albicans and is synthesized from alcohol in the liver.

Acetylcholine: A neurotransmitter manufactured in the brain, used for memory and control of sensory input and muscular output signals.

Acid: Any compound capable of releasing a hydrogen ion; it will have a pH of less than 7.

Acute: Extremely sharp or severe, as in pain, but can also refer to an illness or reaction that is sudden and intense.

Adaptation: Ability of an organism to integrate new elements into its environment.

Addiction: A dependent state characterized by cravings for a particular substance if that substance is withdrawn.

Additive: A substance added in small amounts to foods to alter the food in some way.

Adrenaline: Trademark for preparations of epinephrine, which is a hormone secreted by the adrenal gland. It is used sublingually and by injection to stop allergic reactions.

Aldehyde: A class of organic compounds obtained by oxidation of alcohol. Formaldehyde and acetaldehyde are members of this class of compounds.

Alkaline: Basic, or any substance that accepts a hydrogen ion; its pH will be greater than 7.

Allergenic: Causing or producing an allergic reaction.

Allergen: Any organic or inorganic substance from one's surroundings or from within the body itself that causes an allergic response in an individual is called an allergen. An allergen can cause an IgE antibody mediated or non-IgE mediated response. Some of the commonly known allergens are: pollens, molds, animal dander, food and drinks, chemicals of different kind like the ones found in food, water, inside and outside air, fabrics, cleaning agents, environmental materials, detergent, cosmetics, perfumes, etc., body secretions, bacteria, virus, synthetic materials, fumes of any sort, including pesticide fumes, fumes from cooking, etc., and smog. Emotional unpleasant thoughts like anger, frustration, etc. can also become allergens and cause allergic reactions in people.

Allergic reaction: Adverse, varied symptoms, unique to each person, resulting from the body's response to exposure to allergens.

Allergy: Attacks by the immune system on harmless or even useful things entering the body. Abnormal responses to substances that are usually well tolerated by most people.

Amino acid: An organic acid that contains an amino (ammonia-like NH3) chemical group; the building blocks that make up all proteins.

Anaphylactic shock: Also known as anaphylaxis. Usually happens suddenly when exposed to a highly allergic item. But sometimes, it can also happen as a cumulative reaction. (first two doses of penicillin may not trigger a severe reaction, but the third or fourth could produce an anaphylaxis in some people). An anaphylaxis (this life threatening allergic reaction) is characterized by: an immediate allergic reaction that can cause difficulty in breathing, light headedness, fainting, sensation of chills, internal cold, severe heart palpitation or irregular heart beats, pallor, eyes rolling, poor mental clarity, tremors, internal shaking, extreme fear, angio neurotic edema, throat swelling, drop in blood pressure, nausea, vomiting, diarrhea, swelling anywhere in the body, redness and hives, fever, delirium, unresponsiveness, or sometimes even death.

Antibody: A protein molecule produced in the body by lymphocytes in response to a perceived harmful foreign or abnormal substance as a defense mechanism to protect the body.

Antigen: Any substance recognized by the immune system that causes the body to produce antibodies; also refers to a concentrated solution of an allergen.

Antihistamine: A chemical that blocks the reaction of histamine that is released by the mast cells and basophils during an allergic reaction. Any substance that slows oxidation, prevents damage from free radicals and results in oxygen sparing.

Assimilate: To incorporate into a system of the body; to transform nutrients into living tissue.

Autoimmune: A condition resulting when the body makes antibodies against its own tissues or fluid. The immune system attacks the body it inhabits, which causes damage or alteration of cell function.

Binder: A substance added to tablets to help hold them together.

Blood brain barrier: A cellular barrier that prevents certain chemicals from passing from the blood to the brain.

Buffer: A substance that minimizes changes in pH (Acidity or alkanity).

Candida albicans: A genus of yeast like fungi normally found in the body. It can multiply and cause infections, allergic reactions or toxicity.

Candidiasis: An overgrowth of Candida organisms, which are part of the normal flora of the mouth, skin, intestines and vagina.

Carbohydrate, complex: A large molecule consisting of simple sugars linked together, found in whole grains, vegetables, and fruits. This metabolizes more slowly into glucose than refined carbohydrate.

Carbohydrate, refined: A molecule of sugar that metabolizes quickly to glucose. Refined white sugar, white rice, white flour are some of the examples.

Catalyst: A chemical that speeds up a chemical reaction without being consumed or permanently affected in the process.

Cerebral allergy: Mental dysfunction caused by sensitivity to foods, chemicals, environmental substances, or other substances like work materials etc.

Chronic: Of long duration.

Chronic fatigue syndrome: A syndrome of multiple symptoms most commonly associated with fatigue and reduced energy or no energy.

Crohn's disease: An intestinal disorder associated with irritable bowel syndrome, inflammation of the bowels and colitis.

Cumulative reaction: A type of reaction caused by an accumulation of allergens in the body.

Cytokine Immune system's second line of defense. Examples of cytokines are interleukin 2 and gamma interferon.

Desensitization: The process of building up body tolerance to allergens by the use of extracts of the allergenic substance.

Detoxification: A variety of methods used to reduce toxic materials accumulated in body tissues.

Digestive tract: Includes the salivary glands, mouth, esophagus, stomach, small intestine, portions of the liver, pancreas, and large intestine.

Disorder: A disturbance of regular or normal functions.

Dust: Dust particles from various sources irritate sensitive individual causing different respiratory problems like asthma, bronchitis, hay-fever like symptoms, sinusitis, and cough.

Dust mites: Microscopic insects that live in dusty areas, pillows, blankets, bedding, carpets, upholstered furniture, drapes, corners of the houses where people neglect to clean regularly.

Eczema: An inflammatory process of the skin resulting from skin allergies causing dry, itchy, crusty, scaly, weepy, blisters or eruptions on the skin. skin rash frequently caused by allergy.

Edema: Excess fluid accumulation in tissue spaces. It could be localized or generalized.

Electromagnetic: Refers to emissions and interactions of both electric and magnetic components. Magnetism arising from electric charge in motion. This has a definite amount of energy.

Elimination diet: A diet in which common allergenic foods and those suspected of causing allergic symptoms have been temporarily eliminated.

Endocrine: refers to ductless glands that manufacture and secrete hormones into the blood stream or extracellular fluids.

Endocrine system: Thyroid, parathyroid, pituitary, hypo-thalamus, adrenal glands, pineal gland, gonads, the intestinal tract, kidneys, liver, and placenta.

Endogenous: Originating from or due to internal causes.

Environment: A total of circumstances and/or surroundings in which an organism exists. May be a combination of internal or external influences that can affect an individual.

Environmental illness: A complex set of symptoms caused by adverse reactions of the body to external and internal environments.

Enzyme: A substance, usually protein in nature and formed in living cells, which starts or stops biochemical reactions.

Eosinophil: A type of white blood cell. Eosinophil levels may be high in some cases of allergy or parasitic infestation.

Exogenous: Originating from or due to external causes.

Extract: Treatment dilution of an antigen used in immunotherapy, such as food, chemical, or pollen extract.

Fibromyalgia: An immune complex disorder causing general body aches, muscle aches, and general fatigue.

"Fight" or "flight": The activation of the sympathetic branch of the autonomic nervous system, preparing the body to meet a threat or challenge.

Food addiction: A person becomes dependent on a particular allergenic food and must keep eating it regularly in order to prevent withdrawal symptoms.

Food grouping: A grouping of foods according to their botanical or biological characteristics.

Free radical: A substance with unpaired electrode, which is attracted to cell membranes and enzymes where it binds and causes damage.

Gastrointestinal: Relating both to stomach and intestines.

Heparin: A substance released during allergic reaction. Heparin has antiinflammatory action in the body.

Histamine: A body substance released by mast cells and basophils during allergic reactions, which precipitates allergic symptoms.

Holistic: Refers to the idea that health and wellness depend on a balance between the physical (structural) aspects, physiological (chemical, nutritional, functional) aspects, emotional and spiritual aspects of a person.

Homeopathic: Refers to giving minute amounts of remedies that in massive doses would produce effects similar to the condition being treated.

Homeostasis: A state of perfect balance in the organism, also called "Yin-yang" balance. The balance of functions and chemical composition within an organism that results from the actions of regulatory systems.

Hormone: A chemical substance that is produced in the body, secreted into body fluids, and is transported to other organs, where it produces a specific effect on metabolism.

Hydrocarbon: A chemical compound that contains only hy-drogen and carbon.

Hypersensitivity: An acquired reactivity to an antigen that can result in bodily damage upon subsequent exposure to that particular antigen.

Hyperthyroidism: A condition resulting from over-function of the thyroid gland.

Hypoallergenic: Refers to products formulated to contain the minimum possible allergens and some people with few allergies can tolerate them well. Severely allergic people can still react to these items.

Hypothyroidism: A condition resulting from under-function of the thyroid gland.

IgA: Immunoglobulin A, an antibody found in secretions associated with mucous membranes.

IgD: Immunoglobulin D, an antibody found on the surface of B-cells.

IgE: Immunoglobulin E, an antibody responsible for immediate hypersensitivity and skin reactions.

IgG: Immunoglobulin G, also known as gammaglobulin, the major antibody in the blood that protects against bacteria and viruses.

IgM: Immunoglobulin M, the first antibody to appear during an immune response.

Immune system: The body's defense system, composed of specialized cells, organs, and body fluids. It has the ability to locate, neutralize, metabolize and eliminate unwanted or foreign substances.

Immunocompromised: A person whose immune system has been damaged or stressed and is not functioning properly.

Immunity: Inherited, acquired, or induced state of being, able to resist a particular antigen by producing antibodies to counteract it. A unique mechanism of the organism to protect and maintain its body against adversity by its surroundings.

Inflammation: The reaction of tissues to injury from trauma, infection, or irritating substances. Affected tissue can be hot, reddened, swollen, and tender.

Inhalant: Any airborne substance small enough to be inhaled into the lungs; eg., pollen, dust, mold, animal danders, perfume, smoke, and smell from chemical compounds.

Intolerance: Inability of an organism to utilize a substance.

Intracellular: Situated within a cell or cells.

Intradermal: method of testing in which a measured amount of antigen is injected between the top layers of the skin.

Ion: An atom that has lost or gained an electron and thus carries an electric charge.

Kinesiology: Science of movement of the muscles.

Latent: Concealed or inactive.

Leukocytes: White blood cells.

Lipids: Fats and oils that are insoluble in water. Oils are liquids in room temperature and fats are solid.

Lymph: A clear, watery, alkaline body fluid found in the lymph vessels and tissue spaces. Contains mostly white blood cells.

Lymphocyte: A type of white blood cell, usually classified as T-or B-cells.

Macrophage: A white blood cell that kills and ingests microorganisms and other body cells.

Masking: Suppression of symptoms due to frequent exposure to a substance to which a person is sensitive.

Mast cells: Large cells containing histamine, found in mucous membranes and skin cells. The histamine in these cells are released during certain allergic reactions.

Mediated: Serving as the vehicle to bring about a phenomenon, eg., an IgE-mediated reaction is one in which IgE changes cause the symptoms and the reaction to proceed.

Membrane: A thin sheet or layer of pliable tissue that lines a cavity, connects two structures, selective barrier.

Metabolism: Complex chemical and electrical processes in living cells by which energy is produced and life is maintained. New material is assimilated for growth, repair, and replacement of tissues. Waste products are excreted.

Migraine: A condition marked by recurrent severe headaches often on one side of the head, often accompanied by nausea, vomiting, and light aura. These headaches are frequently attributed to food allergy.

Mineral: An inorganic substance. The major minerals in the body are calcium, phosphorus, potassium, sulfur, sodium, chloride, and magnesium.

Mucous membranes: Moist tissues forming the lining of body cavities that have an external opening, such as the respiratory, digestive, and urinary tracts.

Muscle Response Testing (MRT) or Neuromuscular testing (NST): A testing technique based on kinesiology to test allergies by comparing the strength of a muscle or a group of muscles in the presence and absence of the allergen.

NAET: (Nambudripad's Allergy Elimination Techniques): A technique to permanently eliminate allergies towards the treated allergens. Developed by Dr. Devi S. Nambudripad and practiced by more than 8,000 medical practitioners worldwide. This technique is natural, non-invasive, and drug-free. It has been effectively used in treating all types of allergies and problems arising from allergies. It is taught by Dr. Nambudripad in Buena Park, California. to currently licensed medical practitioners. If you are a licensed medical practitioner, interested in learning more about NAET, or NAET seminars, please visit the website: www.naet.com.

Nervous system: A network made up of nerve cells, the brain, and the spinal cord, which regulates and coordinates body activities.

NST: Neuromuscular testing (NST): A testing technique based on kinesiology to test allergies by comparing the strength of a muscle or a group of muscles in the presence and absence of the allergen.

NTT: A series of standard diagnostic tests used by NAET practitioners to detect allergies is called "Nambudripad's Testing Techniques"or NTT.

Neurotransmitter: A molecule that transmits electrical and/or chemical messages from nerve cell (neuron) to nerve cell or from nerve cell to muscle, secretory, or organ cells.

Nutrients: Vitamins, minerals, amino acids, fatty acids, and sugar (glucose), which are the raw materials needed by the body to provide energy, effect repairs, and maintain functions.

Organic foods: Foods grown in soil free of chemical fertilizers, and without pesticides, fungicides and herbicides.

Outgasing: The releasing of volatile chemicals that evaporate slowly and constantly from seemingly stable materials such as plastics, synthetic fibers, or building materials.

Overload: The overpowering of the immune system due to numerous concurrent exposures or to continuous exposure caused by many stresses, including allergens.

Parasite: An organism that depends on another organism (host) for food and shelter, contributing nothing to the survival of the host.

Pathogenic: Capable of causing disease.

Pathology: The scientific study of disease; its cause, processes, structural or functional changes, developments and consequences.

Pathway: The metabolic route used by body systems to facilitate biochemical functions.

Peakflow meter: An inexpensive, valuable tool used in measuring the speed of the air forced out of the lungs and helps to monitor breathing disorders like asthma.

Petrochemical: A chemical derived from petroleum or natural gas.

pH: A scale from 1 to 14 used to measure acidity and alkanity of solutions. A pH of 1-6 is acidic; a pH of 7 is neutral; a pH of 8-14 is alkaline or basic.

Postnasal drip: The leakage of nasal fluids and mucus down into the back of the throat.

Precursor: Anything that precedes another thing or event, such as physiologically inactive substance that is converted into an active substance that is converted into an active enzyme, vitamin, or hormone.

Prostaglandin: A group of unsaturated, modified fatty acids with regulatory functions.

Radiation: The process of emission, transmission, and absorption of any type of waves or particles of energy, such as light, radio, ultraviolet or X-rays.

Receptor: Special protein structures on cells where hormones, neurotransmitters, and enzymes attach to the cell surface.

Respiratory system: The system that begins with the nostrils and extends through the nose to the back of the throat and into the larynx and lungs.

Rotation diet: A diet in which a particular food and other foods in the same "family" are eaten only once every four to seven days.

Sensitivity: An adaptive state in which a person develops a group of adverse symptoms to the environment, either internal or external. Generally refers to non-IgE reactions.

Serotonin: A constituent of blood platelets and other organs that is released during allergic reactions. It also functions as a neurotransmitter in the body.

Sublingual: Under the tongue–method of testing or treatment in which a measured amount of an antigen or extract is administered under the tongue, behind the teeth. Absorption of the substance is rapid in this way.

Supplement: Nutrient material taken in addition to food in order to satisfy extra demands, effect repair, and prevent degeneration of body systems.

Susceptibility: An alternative term used to describe sensitivity.

Symptoms: A recognizable change in a person's physical or mental state, that is different from normal function, sensation, or appearance and may indicate a disorder or disease.

Syndrome: A group of symptoms or signs that, occurring together, produce a pattern typical of a particular disorder.

Synthetic: Made in a laboratory; not normally produced in nature, or may be a copy of a substance made in nature.

Systemic: Affecting the entire body.

Target organ: The particular organ or system in an individual that will be affected most often by allergic reactions to varying substances.

Toxicity: A poisonous, irritating, or injurious effect resulting when a person ingests or produces a substance in excess of his or her tolerance threshold.

Freedom From Environmental Sensitivities

Resources

www.naet.com - The NAET website for
information regarding NAET

Nambudripad Allergy Research Foundation (NARF)
6714 Beach Blvd.
Buena Park, CA 90621
(714) 523-0800
A Nonprofit foundation dedicated to NAET research

NAET Seminars
6714 Beach Blvd.
Buena Park, CA 90621
(714) 523-8900
NAET Seminar information

Delta Publishing Company (for Books on NAET)
6714 Beach Blvd.
Buena Park, CA 90621
(714) 523-0800
E-mail: naet@earthlink.net

Jacob Teitelbaum MD
CFS/Fibromyalgia Therapies
Author of the best selling book:

"From Fatigued to Fantastic!" and
"Three Steps to Happiness! Healing Through Joy"
(410) 573-5389
www.EndFatigue.com

Environmentally Safe Products

Quantum Wellness Center

Drs. Dave & Steven Popkin

1261 South Pine Island Rd.

Plantation, FL 33324

(954) 370-1900/ Fax: (954) 476-6281

E-mail: buddha327@aol.com

Cotton Gloves and other Environmentally

Safe Health Products

Janice Corporation

198 US Highway 46

Budd Lake, NJ 07828-3001

(800) 526-4237

Herbal Supplements

Kenshin Trading Corporation
1815 West 213th Street, Ste. 180
Torrance, CA 90501
(310) 212-3199

Phenolics

Frances Taylor/Dr. Jacqueline Krohn
Los Alamos Medical Center, Ste.136
3917 West Road
Los Alamos, NM 87544
(505) 662-9620

Enzyme Formulations, Inc
6421 Enterprise Lane
Madison, WI 53719
(800) 614-4400

Allergies Lifestyle & Health
205 Center Street, Ste. B.
Eatonville, WA 98328
(360) 832-0858
Health Products

Bio Meridian
12411 S. 265 W. Ste. F
Draper, UT 84020
(801) 501-7517
Computerized Allergy Testing Services

Star Tech Health Services, LLC
1219 South 1840 West
Orem, Utah 84058
(888) 229-1114
Computerized Allergy Testing Services

Thorne Research Inc.
P.O. Box 25
Denver, ID 83825
(208) 263-1337
Herbs and Vitamins

Earth Calm
3805 Windermere Lane
Oroville, CA 95965
(530) 534 9982

Dreamous Corporation
12016 Wilshire Blvd. # 8
Los Angeles, CA 90025
(310) 442 8544

K & T Books
LAMC, Ste. 136,
3917 West Road
Los Alamos, NM 87544
(505) 662 9620

Neuropathways EEG Imaging
427 North Canon Dr. # 209
Beverly Hills, CA 90210
(310) 276 9181

CHI/KHT
P.O. Box 5309
Hemet, CA 92544
(909) 766 1426
Health Products

Biochemical Laboratories
P.O. Box 157
Edgewood, NM 87015
(800) 545 6562

Green Healing Center C
1700 Sansom St., Ste.800
Philadelphia, PA 19103
215-751-9833

Bibliography

Arbuckle, B.E. Cranial birth injuries. Academy of Applied Osteopathy, *yearbook 1945:63.*

Arbuckle, B.E. Early cranial considerations. *JAOA 48(2):315-320. Arbuckle, B.E.* Effects of uterine forceps upon the fetus. *JAOA 54(5)#9:499-508.*

Asperger, H. (1979). *Problems of Infantile Autism.* Communication. 13, 45-52.

Autism Autoimmunity Project, *"The causes of autism and the need for immunological research: Excerpts from the autism literature,"* available on the internet at //http/libnt2.lib.tcu.edu/staff/lruede/ immresearch.html (Autism Autoimmunity Project, www.gri.net/ truegrit) "U.S. Officials investigate 'cluster' of autism in New Jersey Town," CNN, February 1, 1999.

Baron, Judy and Sean Barron. *There's a Boy in Here.* New York: Simon and Shuster, 1993.

Baron-Cohen, Simon, and Patrick Bolton. *Autism: The Facts.* Oxford: Oxford University Press, 1993.

Bauman, M.L., R.A. Filipek and T.L. Kemper. 1997. Early infantile autism. *International review of neurobiology 41:367-386.*

Bettelheim, B. *The Empty Fortress: Infantile Autism and the Birth of Self.* New York: Free Press, 1967.

Betts, Carolyn. *A Special Kind of Normal.* New York Scribner, 1983.

Brandl, Cherlene. *Facilitated Communication: Case Studies*-- See Us Smart! Ann Arbor Maine: Robbie Dean Press, 1999.

Bristol, M., D.J. Cohen et al. 1996. *State of science of autism:* report to the National Institutes of Health. Journal of autism and developmental disorders 26(2):121-154.

Callahan, Mary. *Fighting for Tony.* New York: Simon and Schuster, 1997.

Capps, L., Sigman, M., and P. Mundy. *Attachment Security in Children with Autism. Development and Psychopathology,* 6, 24999-261.

Castleman, *Michael. Nature's Cures.* Emmaus, PA: Rodale Press, 1996.

Chopra, Deepak. *Perfect Health---The Complete Mind/Body Guide.* New York: Harmony Books, 1991.

Chugani, D., O. Musik et al. 1997. *Altered serotonin synthesis in the dentatothalamococrtical pathway in autistic boys.* Annals of Neurology 42(10)#4:666-669.

Cohen, D.J., Donnellan, A. and R. Paul (eds). *Handbook of Autism and Pervasive Development Disorders.* New York: Wiley, 1987.

Dawson, G., ed. 1989. *Autism: nature, diagnosis and treatment.* New York. Guilford Press 23A. Wales, pers. comm. 1999. My own pers. experience.

Cook, E.H., Jr., 1996. *Brief report: pathophysiology of autism:* neurochemistry. Journal of autism and developmental disorders 26(2):221-225.

Courchesne, E., J. Townsend, et al. 1994. *The brain in infantile autism: posterior fossa structures are abnormal.* Neurology 44:214-223.

Courchesne, E., R. Yeung-Courchesne, et al. 1988. *Hypoplasis of cerebellar lobules VI-VII in infant autism.* New England Journal of Medicine 318:1349-1354.

Courchesne, E. 1999. Correspondence re: *an MRI study of autism: the cerebellum revisited.* Neurology 52:1106.

Diagnostic and Statistical Manual of Mental Disorders. (4th ed.). Washington, D.C: American Psychiatric Association, 1994.

Diagnostic and Statistical Manual of Mental Disorders. DSM IV-TR, *(4th ed.).* Washington, D.C: American Psychiatric Association, 2000.

DeLong, G.R. 1999. *Autism: new data suggesting new hypothesis:* views and reviews. Neurology 52:911-916.

Doctors warn developmental disabilities epidemic from toxins, LDA (Learning Disabilities association of America) Newsbriefs 35.4 (July/August 2000): 3; executive summary from the report by the Greater Boston Physicians for Social Responsiblity, In *"Harm's way–Toxic Threats to Child development,"* available at www.igc.org/psr/ihw.htm;for LDA, www.ldanatl.org.

Dillon, Katleen M. Living with Autism: *The Parents' Stories. Boone,* NC: Parkway, 1995.

Edelson, stephen, Ph.D., Rimland, Bernard, Ph.D. *Treating Autism,* Autism Research Institute, San Diego, 2003.

Ernst, M., A.J. Zametkin, et al. 1997. *Low medical prefrontal dopaminergic activity in autistic children.* The Lancet 350(8):638.

Firth, U. *Autism and Asperger Syndrome.* Cambridge, England: University Press, 1991.

Frymann, V.M. *Relation of disturbances of craniosacral mechanisms to symptomalogy of the newborn:* a study of 1,250 infants. JAOA 65(10):1059-1075.

Fryman, V.M., (1976) *Trauma of birth.* Osteopathic Annals 4(22):8-14.

Frymann, V.M., R.E. Carney, et al. *Effect of osteopathic medical management on neurologic development in children.* JAOA 92(6):729-744.

Frymann, V.M. *Learning difficulties of children viewed in the light of osteopathic concept.* JAOA 76(1):46-61.

Garrison, William. Small Bargains: *Children in Crisis and the Meaning of Parental Love.* New York: Simon and Schuster, 1993.

Gerlach, Elizabeth K. *Autism Treatment Guide.* Four-Leaf Press, 1993.

Grandin, Temple, and Margaret M. Scariano. *Emergence: Labeled Autistic.* Ney York: Warner Books, 1996.

Harrington, Kathie. For Parents and Professionals: *Autism. Lingui Systems*, 1998.

Jealous, J.1997. Conservations: *healing and the natural world.* Alternative therapies 3(1):68-75.

Kane, P. 1997. *Peroxisomal Disturbances in Autistic Spectrum Disorder.* Journal of Orthomolecular Medicine 12 (4):207-218.

Kephart, Beth. A Slant of the Sun: *One Child's Courage.* W.W. Norton: 1998.

Kaufman, Barry Neil. *Son-Rise.* New York: Harper and Row, 1976.

Kaufman, Barry Neil and Samarhia Lyte Kaufman. *Son-Rise: The Miracle Continues.* Kramer, 1994.

Landrigan, P., and J. Witte. *Neurologic Disorders Following Measles Virus Vaccinations.* JAMA233: 1459 (1973).

Lawrence Lavine, *"Osteopathic and alternative medicine aspects of autistic spectrum disorders,"* article on the internet (available at trainland.tripod.com/lawrencelavine.htm).

Lippincott, R.C. *Cranial Osteopathy.* AAO Year Book 1947:103-111.

Lynne Cannon, *"The Environment and Learning Disabilities,"* LDA Newsbriefs 35:4 (July/August 2000): 1; forLDA, www.ldanatl.org.

Lynne Cannon, *"The Environment and Learning Disabilities,"* LDA Newsbriefs 35:4 (July/August 2000): 1; for LDA, www.ldanatl.org.

Madaule, Paul. When Listening Comes Alive: A Guide to Effective Learning and communication, Norval, Ontario:Moulin, 1994.

Manning, Anita. 1999. *Vaccine-autism link feared. USA Today*, 16 Aug. 99.

Matson, J.L., D.A. Benavidez et al. 1996. *Behavioral treatment of autistic persons:* a review of research from 1980 to the present. Research in developmental disabilities 17(6):433-465.

McBean, Eleanor. *Vaccinations Do Not Protect.* Manachaea, TX: Health Excellence Systems, 1991.

Bibliography

McGilvery, Robert W., and Gerald W. Goldstein. *Biochemistry---A Functional Approcah*. Philllladphia, PA: W.B. Saunders Company, 1983.

Miller, Neil Z. Vaccines: *Are They Really Safe and Effective?* Santa Fe, New Mexico: New Atlantian Press, 1992.

Nambudripad, Devi S., M.D., D.C., L.Ac., Ph.D. *Say Good-bye to Illness,* English, 3rd. Ed., Buena Park, California, Delta Publishing, 2002.

Nambudripad, Devi S., M.D., D.C., L.Ac., Ph.D. *Say Good-bye to Illness,* French Edition, Buena Park, California, Delta Publishing, 2001.

Nambudripad, Devi S., M.D., D.C., L.Ac., Ph.D. *Say Good-bye to Illness,* Spanish Edition, Buena Park, California, Delta Publishing, 1999.

Nambudripad, Devi S., M.D., D.C., L.Ac., Ph.D. *The NAET Guide Book, 6th Edition,* Buena Park, California, Delta Publishing, 2004.

Nambudripad, Devi S., M.D., D.C., L.Ac., Ph.D. *The NAET Guide Book, French Edition,* Buena Park, California, Delta Publishing, 2003.

Nambudripad, Devi S., M.D., D.C., L.Ac., Ph.D. *The NAET Guide Book, German Edition,* Buena Park, California, Delta Publishing, 2004.

Nambudripad, Devi S., M.D., D.C., L.Ac., Ph.D. *Say Good-bye to Allergy-related Autism,* 2nd. Ed., Buena Park, California, Delta Publishing, 2004.

Nambudripad, Devi S., D.C., L.Ac., Ph.D. *Say Good-bye to ADD and ADHD,* Buena Park, California, Delta Publishing, 1999.

Nambudripad, Devi S., M.D., D.C., L.Ac., Ph.D. *Say Good-bye to Your Allergies,* Buena Park, California, Delta Publishing, 2004.

Nambudripad, Devi S., D.C., L.Ac., Ph.D. *Say Good-bye to Children's Allergies.*, Buena Park, California, Delta Publishing, 1999.

Nambudripad, Devi S., D.C., L.Ac., Ph.D. *Living Pain Free.*, A Self-help Book on Acupressure Therapy, Buena Park, California, Delta Publishing, 1997.

Nambudripad, Devi S., M.D., D.C., L.Ac., Ph.D. *Say Good-bye to Your Environmental Sensitivities,* Buena Park, California, Delta Publishing, 2004.

Freedom From Environmental Sensitivities

Nambudripad, Devi S., M.D., D.C., L.Ac., Ph.D. *Say Good-bye to Your Chemical Sensitivities,* Buena Park, California, Delta Publishing, 2004.

Nambudripad, Devi S., M.D., D.C., L.Ac., Ph.D. *Say Good-bye to Asthma,* Buena Park, California, Delta Publishing, 2004.

Oppenheim, Rosalind. *Effective Teaching Methods for Autistic Children.* Springfield, Illinois: Charles C. Thomas, 1974.

Orange County register, *Newsbriefs: Focus/Health,* November 19, 2003.

Pangborn, Jon, B. PhD., and Sidney Baker, M.D. *Biomedical Assessment Options For Children With Autism And Related problems*, Autism research Institute, San Diego, 2002.

Piaget, J. *The Construction of Reality in the Child.* New York: W.W. Norton, 1962.

Piven, J., E. Nehme, et al. 1992. *Magnetic resonance imaging in autism: measurement of the cerebellum, pons, and fourth ventricle.* Biologic Psychiatry 31:491-504.

Rutter, M. Autistic Children: *Infancy to Adulthood. Seminars in Psychiatry and Allied Disciplines,* 24, 513-531.

Rapp, Doris. *Is This Your Child?* New York: William Morrow and Company, 1991.

Rapp, Doris. *Our Toxic World: A Wake Up Call. Environmental Medical Research Foundation, Buffalo, NY.. Tel. 1-800-787-8780.* Website: www.drrapp.com.

Rapin, I., and R. Katzman, 1998. *Neurobiology of autism.* Annals of neurology 43(1):7-14.

Rapin, I. 1999. *Autism in search of a home in the brain.* Neurology 52:902-904.

Rea, William J. *Chemical Sensitivity.* Boca Raton, FL: Lewis Publishers, 1996.

Richard Leviton, *"The Healthy Living Space,"* Charlotteville, Virginia: Hampton Roads, 2001:2.

Richard Leviton, *"The Healthy Living Space,"* Charlotteville, Virginia: Hampton Roads, 2001:3.

Rimland, Bernard. *Controversies in the Treatment of Autistic Children: Vitamin and Drug Therapy,* J. Child Neurol. 3 Suppl: S6-16, 1988.

Rimland, Bernard. *Secretin Update.* Autism Research Review International. March, 1999.

Rimland, Bernard. Vaccinations: *The Overlooked Factors. Autism* Research Review International, 1998.

Rimland, Bernard. *Candida-Caused Autism?* Autism Research Review International, 1988.

Rimland, Bernard, Ph.D., in *"Defeat Autism Now (*DAN!) Mercury Detoxification Consensus Group Position Paper," Autism research institute, San Diego, California, May 2001:3.

Sanua, VD. *Studies in Infantile Autism.* Child Psychiatry Hum. Dev. 19(3):207-27, 1989.

Smalley, S.L., Asarnow, R.F. and A. Spence. (1988) *Autism and Genetics:* A Decade of Research. Archives of General Psychiatry, 455, 953-961.

Smalley, S.L. and F. Collins. 1996. *Brief report: genetic, prenatal and immunologic factors.* Journal of autism and developmental disorders 26(2):195-198.

Smith, M.D. *Autism and Life in the Community:* Successful Interventions for Behavioral Challenges. Paul Brooks, 1990.

Smith, CW, Electromagnetic Man: *Health and Hazard in the Electrical Environment*, Martin's Press, 1989, 90, 97.

Smith CW, Environmental Medicine: *Electromagnetic Aspects of Biological Cycles,* 1995:9(3):113-118.

Smith CW., Electrical Environmental Influences on the Autonomic Nervous System, 11th. Intl. Symp. on *"Man and His Environment in Health and Disease,"* Dallas, Texas, February 25-28, 1993.

Smith CW., Electromagnetic Fields and the Endocrine System, 10th. Intl. Symp. on *"Man and His Environment in Health and Disease,"* Dallas, Texas, February 27- March 1, 1992.

Smith CW., Basic Bioelectricity: Bioelectricity and Environmental Medicine, 15th. Intl. Symp., on *"Man and His Environment in Health and Disease,"* Dallas, Texas, February 20-23, 1997. (Audio Tapes from: Professional Audio Recording, 2300 Foothill Blvd. #409, La Verne, CA.

Speer, F., ed. 1970. *Allergy of the nervous system.* Springfield: Charles C. Thomas pub.

Sutherland, W.G. Bent Twigs: *compression of the condylar parts of the occiput.* Teachings in the science of osteopathy. Ed. A.L. Wales. Rurda Press, 1990, 107-117.

Sutherland, W.G. (1943) *The Cranial Bowl.* JAOA 48(4):348-53.

Sutherland, A.S., and A.L. Wales, eds. 1967. *Contributions of thought*: collected writings of William Gamer Sutherland 1914-1954. The Sutherland Cranial Teaching Foundation.

Sutherland, W.G. (1939) *The cranial bowl:* a treatise relating to cranial articular mobility, cranial articular lesions, and cranial technique.Free Press, 1994.

State of California. California Health and Human Services Agency. Dept of Developmental Services. *Changes in the population of persons with Autism/PDD in California's Developmental Services* System:1987 through 1998. A report to the Legislature: March 1, 1999.

Stehl, Annabel. *The Sound of a Miracle, A Child's Triumph Over Autism.* Doubleday, 1991.

Stehl, Annabel. Dancing in the Rain: *Stories of Exceptional Progress by arents of Children with Special Needs* Georgiana, 1995.

Strom, Charles M. Heredity and Ability: *How Genetics Affects Your Child and What You Can Do About It* NewYork: Plenum Press, 1990.

Susser, M. 1973. *Causal thinking in the health sciences: conceptsand strategies of epidemiology.* 3rd ed. New York: Oxford UP.

Sui, Choa Kok, Pranic Healing, Samuel Wiser, 1990.

"The Holistic Physician–Autism," Alternative Medicine Digest 14 (September 1996):20.

The journal of energetics and Complementary Medicine, Spring 2005.

Teitlebaum, Jacob, M.D., *From Fatigued to Fantastic,* 1st ed., 1996, 2nd. ed., Avery Penguin Putnam, 2001.

Tager-Flusberg, H. *Sentence Comprehension in Autistic Children.* Applied Psycholinguistics, 2, 5-24.

U.S. Department of Health and Human services, *"Mental health: A report of the Surgeon General,"* Rockville, Maryland: U.S. Department of Health and Human Services, Substance Abuse and Mental Health Services Administration, Center for Mental health services, National Institute of Health, National Institute of Mental Health, 1999.

Volkmar, F.R. and D.J. Cohen. (1991). *Debate and Argument: The Ultility of the Term Pervasive Developmental Disorder.* Journal of Child psychology and Pscyhiatry. 32, 1171-1172.

Volkmar, F.R., Paul R., and D. Cohen. (1985). *The Use of "Asperger's Syndrome."* Journal of Autism and Developmental Disorders, 15, 437-439.

Wales, A.L. *Cranial diagnosis.* Journal of the Osteopathic Cranial Association 1948:14-23.

Weil, Andrew. *Health and Healing*---Understanding Conventional and Alternative Medicine. Dorling Kindersley, 1995.

Wild, Gaynor, and Edward c. Benzel *Essentials of Neurochemistry.* Boston, MA: Joues and Bartlett Publishers, 1994.

Williams, Donna. *Nobody Nowhere,* Random House, 1992.

Wing, Lorna. *Early Childhood Autism.* Oxford: Pergamon Press, 1976.

Wing, Lorna. *Autistic Children:* A Guide for Parents and Professionals, 2nd edition. New York: Brunner/ Mazel, 1985.

Weiss, Jordan, M.D., *Psychoenergetics,* 2nd. ed., Oceanview Publishing, 1995.

Woods, R.H. 1973. *Structural normalization in infants and children with particular references to disturbances of the central nervous system.* JAOA 72(5):903-08.

Zong, Linda, *Chinese Internal Medicine, lectures* at SAMRA University, Los Angeles, 1985.

Case Histories from the Author's private practice,1984-present.

Freedom From Environmental Sensitivities

Index